The Tactical Guide to
SIX SIGMA
Implementation

SURESH PATEL

CRC Press
Taylor & Francis Group
Boca Raton   London   New York

CRC Press is an imprint of the
Taylor & Francis Group, an **informa** business

A PRODUCTIVITY PRESS BOOK

CRC Press
Taylor & Francis Group
6000 Broken Sound Parkway NW, Suite 300
Boca Raton, FL 33487-2742

© 2016 by Taylor & Francis Group, LLC
CRC Press is an imprint of Taylor & Francis Group, an Informa business

International Standard Book Number-13: 978-1-4987-4538-3 (Hardback)

Library of Congress Cataloging-in-Publication Data

Names: Patel, Suresh (Quality management consultant), author.
Title: The tactical guide to six sigma implementation / Suresh Patel.
Description: Boca Raton, FL : CRC Press, 2016. | Includes bibliographical
references and index.
Identifiers: LCCN 2015046413 | ISBN 9781498745383
Subjects: LCSH: Six sigma (Quality control standard) | Quality
control--Management. | Total quality management.
Classification: LCC TS156.17.S59 P37 2016 | DDC 658.4/013--dc23
LC record available at http://lccn.loc.gov/2015046413

**Visit the Taylor & Francis Web site at
http://www.taylorandfrancis.com**

**and the CRC Press Web site at
http://www.crcpress.com**

I dedicate this book series to K.K. Nair, executive director of the Ahmedabad Management Association (AMA), who encouraged me to write this book after observing excellent feedback from the delegates at the first-ever Lean Six Sigma three-day course at the AMA on June 23–25, 2011.

And also to my dear wife, Pushpa, who had to bear many disruptions and inconveniences without my help and without whose full cooperation this book would not have materialized.

Contents

List of Figures xi
List of Tables xvii
Foreword xix
Preface xxv
Acknowledgments xxvii
Making It Big in Manufacturing Products and Providing Service......xxxi

Chapter 1 Introduction to Six Sigma: A Tactical Strategy.............. 1

Origin of Six Sigma at Motorola..1
Understanding the Variation, Spread, and Sigma 2
DMAIC Process 4

Chapter 2 What Is Six Sigma?... 7

Basic Statistics 7
    Types of Statistics...7
    Analytical Statistics 8
Definition of Sample Statistic, Population,
and Population Parameter..8
Central Limit Theorem and Standard Normal Distribution....9
Central Tendencies 11
Measure of Dispersion 12
    Six Sigma as a Metric..13
        Statistical Interpretation of Six Sigma.......................13
        Why Is the Normal Distribution Useful? 14
        Parts per Million and Sigma Level 15
    Six Sigma as a Management System................................16
    Six Sigma as a Methodology..17

Chapter 3 Business Performance Measures 19

Organizational Processes and Their Impact
on the Organization 19
Defining Owners and Stakeholders 19
Balanced Scorecard 20

Key Performance Indicators...20
Important Measures of Quality Performance 22
    Defects per Unit 22
    Defects per Million Opportunities..................................23
    Throughput Yield..23
    Rolled Throughput Yield..23
    Parts per Million..24
    Cost of Poor Quality..24
        Business Failure Cost Measurements
        in Manufacturing and Finance 27
    Financial Measures...30
        Project Financial Benefits 30
        Financial Benefits Defined..30
        Cost–Benefit Analysis..30
        Return on Assets 32
        Return on Investment...32
        Net Present Value...32
    Team Management...35

Chapter 4 Define...37
Key Concepts and Tools 37
Voice of the Customer...38
    Internal and External Customers....................................38
        Customer–Supplier Chain 38
    Customer Satisfaction 38
    Customer Satisfaction Survey 39
    Quality Function Deployment...40
        Expanded House of Quality 41
        Next Stage..50
    CTQ Flow-Down 51
        Critical to X (CTX) Requirements.................................51
        CTQ Flow-Down Example 51
Project Charter..53
    Project Statement 53
        Business Case...54
    Project Scope 54
    Project Goal..54
        SMART Goals..56

Project Performance..56
   Project Performance Metrics............................57
Project Tracking...60
   Gantt Chart...60
      Gantt Chart Procedure..................................60
   Critical Path Analysis..62
      Critical Path Analysis Procedure................63
   PERT Chart..63
      PERT Chart Example 64
      PERT Chart Key Terms...................................65
      PERT Chart Procedure....................................65
Define Phase Summary..66

Chapter 5 Measure...67

Process Characteristics...67
   Process Handoff Diagram 70
      Challenges at Cross-Functional Areas............72
      Data Collection...72
      Qualitative and Quantitative Data 73
      Measurement Scales...73
Measurement System Analysis 74
   Sampling Strategy..75
   Measurement Equipment 76
   Terms and Definitions...77
      Bias 77
      Precision 79
   Measurement System Process 80
      Step 1: Prepare for the Study 81
      Step 2: Understand and Evaluate Stability....82
      Step 3: Evaluate Resolution 87
      Step 4: Determine Accuracy...........................88
      Step 5: Calibrate the Instrument....................89
      Step 6: Evaluate Linearity..............................89
      Step 7: Determine Repeatability and Reproducibility....91
Process Capability Measurement 98
   Process Capability Assessment 100
   Cp, Cpk, Pp, Ppk 100
   Two Opinions: Process Capability Studies................104

Process Capability Study Procedure 105
Measure Phase Summary..105

Chapter 6   Analyze 107

Regression and Correlation..108
Introduction 108
Variables and Relationships 108
Analysis of Variance..110
Hypothesis Testing...112
Hypothesis Testing with *F*-Statistics............................112
Hypothesis Test Definitions 113
How to Write H_0 and H_a 113
Two Outcomes of the Hypothesis Test 114
Gap Analysis...115
Root Cause Analysis...116
Five Whys...116
Fault Tree Analysis 116

Chapter 7   Improve.. 119

Prioritization through Cause-and-Effect Matrix...............119
Improve Process through Lean Six Sigma 121
Theory of Constraints 121

Chapter 8   Control... 125

Statistical Process Control..125
SPC Background 125
Objectives of SPC.. 126
Uses of SPC Tools.. 126
Benefits of SPC 126
Basic SPC Concepts 126
Control Chart Guide.. 128
Control Chart Road Map...129
Variable Control Charts...130
Variable Data Charts 130
Equations for Variable Data Control Charts 130
Control Chart Constants...130

X-Bar and s Chart 130
X-Bar and s Chart 130
X-MR or I-XR Chart 137
Attribute Data Control Charts......................................140
    p Charts 141
    np Chart...143
    c and *u* Charts...146
Control Plan 154
  Introduction 154
Sustenance of Improvements 156
  Lessons Learned..156
  Implementation of Training Plan...................................156
  Standard Operating Procedures and Work
  Instructions..157
  Ongoing Evaluation..158

Chapter 9   Design for Six Sigma 161
DFSS and SIX SIGMA...161
DFSS and ROI 162
DFSS Methodologies 163
  DMADV...164
  DMADOV...164
  DFSS Teams 165
  Design for X 166
    Concepts and Techniques for DFX...............................166
    Concepts for DFX Diagram..168
  Reliability 168
    Bathtub Curve 170
  Tolerance 177
    Statistical Tolerance 177
    Stack Tolerance 178
    Statistical Tolerancing 180
  Marketing and Porter's Five Forces Analysis...................181
    Marketing..181
    Porter's Five Forces Analysis 182
  TRIZ 183
    TRIZ and DFSS 185

Chapter 10 Case Study: Improvement Using Lean Six Sigma 187

Recognize and Identify Key Business: Game of Cricket.....187
Case Study Learning Objectives...191
Define Phase: Identify Evidence of Singh's Cricketing
Problem 192
10 Steps Leading to the Team Charter.............................192
Practice Drill Process...195
Gary and Singh's Team Charter 195
Measure Phase...196
Critical to Process...196
Analyze Phase 196
Improve Phase..199
Singh's Practice and Training Regime..............................199
Control Phase 200
Standardize...201
Integrate 201

Glossary.. 203

Bibliography... 211

Index.. 213

About the Author.. 223

List of Figures

Figure 1.1 Types of variations and their methods of removal 2

Figure 1.2 Various means of measuring variation 3

Figure 1.3 DMAIC process 5

Figure 1.4 Generic Six Sigma process 5

Figure 2.1 Normal distribution 9

Figure 2.2 Normal distribution is symmetrical 11

Figure 2.3 Mean, median, and mode in normal and skewed
distributions 12

Figure 2.4 Formulas for measures of dispersion 13

Figure 2.5 Standard normal distribution 14

Figure 2.6 Significance of sigma level in defect assessment 15

Figure 2.7 Relation between parts per million and sigma level 16

Figure 3.1 Organizational process with feedback loop 20

Figure 3.2 Balanced scorecard diagram 21

Figure 3.3 Example of a typical balanced scorecard 22

Figure 3.4 Cost of poor quality 27

Figure 3.5 Business failure costs 28

Figure 3.6 Business failure costs analyzed 29

Figure 3.7 Business cost analysis as an improvement driver 29

Figure 4.1 Customer–supplier chain to understand and agree
on mutual requirements 39

Figure 4.2 House of quality 40

Figure 4.3 Expanded house of quality 42

Figure 4.4 Preferred customer requirements 43

Figure 4.5   Competitor rating by the customer 43

Figure 4.6   Determine company's ability to meet customer needs 44

Figure 4.7   Rate organizational difficulty to meet customer requirements 45

Figure 4.8   Technical analysis of competitor products 46

Figure 4.9   Target values for technical specifications 47

Figure 4.10   Correlation matrix 48

Figure 4.11   Absolute and relative importance 49

Figure 4.12   CTQ flow-down 52

Figure 4.13   Project phases, decision gates, and performance review map 57

Figure 4.14   Project charter 58

Figure 4.15   Project status report (4-up chart) 59

Figure 4.16   Gantt chart 61

Figure 4.17   Critical path analysis 62

Figure 4.18   PERT chart 64

Figure 5.1   Quality system documents and their hierarchy 68

Figure 5.2   Spaghetti diagram for a machine shop 70

Figure 5.3   Circle diagram for Six Sigma benefits 71

Figure 5.4   Process handoff diagram 71

Figure 5.5   "Breaking the Barriers" cartoon 72

Figure 5.6   Bias 77

Figure 5.7   Degree of accuracy 78

Figure 5.8   Two components of precision 79

Figure 5.9   Measurement system errors 80

Figure 5.10   Data for stability test 83

Figure 5.11   Calculated UCL and LCL values 85

Figure 5.12   Mean and range charts 86

Figure 5.13 Chart showing out-of-control conditions 86

Figure 5.14 Linearity graph 91

Figure 5.15 Repeatability and reproducibility data sheet.......................95

Figure 5.16 Summary of equations for one standard of deviation
used in the gage R&R study 96

Figure 5.17 Repeatability and reproducibility report 97

Figure 5.18 Basic process capability diagram 100

Figure 5.19 Process capability histograms 101

Figure 6.1 Continuous data classification..108

Figure 6.2 Graph showing typical relationship between two
variables..109

Figure 6.3 Correlation between the gap and the door closing effort...109

Figure 6.4 Common FTA gate symbols...117

Figure 7.1 Cause-and-effect matrix 120

Figure 7.2 Lean tools to improve processes.. 122

Figure 8.1 Common causes. 127

Figure 8.2 Special causes.. 128

Figure 8.3 Control chart selection guide...129

Figure 8.4 Equations for variable control charts 131

Figure 8.5 Location of X-bar and s chart...133

Figure 8.6 X-bar s chart data sheet.. 134

Figure 8.7 X-bar s chart calculation of grand mean 134

Figure 8.8 X-bar and s chart calculation of control limits....................135

Figure 8.9 Draw control limits and grand mean centerline
and plot sample means 136

Figure 8.10 Draw control limits and S-bar centerline; plot sample
S-values...136

Figure 8.11 Completed X-bar and s graph...137

Figure 8.12 Location of X-MR and I-XR charts......................................138

Figure 8.13   Formulas for X-MR and I-XR charts...................................139

Figure 8.14   Data sheet for X-MR or I-XR chart...............................139

Figure 8.15   Draw control limits and mean centerline and plot sample means...140

Figure 8.16   Formulas for attribute data charts...............................141

Figure 8.17   Location of *p* chart..142

Figure 8.18   Data sheet for *p* chart...143

Figure 8.19   Calculation of *p*-bar, *n*-bar, and control limits...............143

Figure 8.20   Drawing of a *p* chart..144

Figure 8.21   Location of *np* chart...144

Figure 8.22   Data sheet to calculate *np*-bar and *p*-bar....................145

Figure 8.23   Calculation of UCL and LCL for *np* chart.....................145

Figure 8.24   Completed *np* chart..146

Figure 8.25   Location of *c* chart..147

Figure 8.26   Data sheet for *c* chart...148

Figure 8.27   Calculation of UCL and LCL for *c* chart.......................148

Figure 8.28   Completed *c* chart..149

Figure 8.29   Location of *u* chart..150

Figure 8.30   Data sheet and formula for *u* chart............................150

Figure 8.31   UCL and LCL data for *u* chart...................................151

Figure 8.32   Plot for *u*-bar centerline and UCL and LCL for $n = 1$ and $n = 2$, respectively...151

Figure 8.33   Completed *u* chart..152

Figure 8.34   Math for *u* chart..153

Figure 8.35   Control plan form...155

Figure 9.1   Venn diagram showing Six Sigma and DFSS functions....162

Figure 9.2   Product life cycle cost..163

Figure 9.3   DMADV process...164

Figure 9.4 DFX concept diagram...169

Figure 9.5 Bathtub curve..170

Figure 9.6 Wear-out period normal distribution. 174

Figure 9.7 Part dimensional specification. 178

Figure 9.8 Stack tolerance. 180

Figure 9.9 Porter's five forces analysis....................................183

Figure 9.10 Problem of psychological inertia.........................184

Figure 10.1 Game of cricket..188

Figure 10.2 Oval-shaped cricket ground with pitch. 189

Figure 10.3 Dimensions of the pitch 189

Figure 10.4 Batsman and wicket keeper with personal protection equipment... 190

Figure 10.5 Names of the directions in which a ball can be hit......... 190

Figure 10.6 Team charter. 195

Figure 10.7 Problem analysis..198

Figure 10.8 Data analysis using Pareto chart.........................199

List of Tables

Table 2.1 Sigma level and DPMO 16

Table 3.1 Four perspectives to determine organizational strategy 21

Table 3.2 Sigma conversion table showing relations
among DPMO, sigma, and Cpk 25

Table 3.3 Sigma, DPMO, yield, and short- and long-term Cpk
comparison 26

Table 3.4 Cash flow analysis table 34

Table 4.1 Eight hows (Company ability to meet customer needs) 50

Table 4.2 Eight whats (Customer requirements) 50

Table 4.3 Example of business case summary 55

Table 4.4 Poor and well-written problem statements 56

Table 5.1 Measurement scales 74

Table 5.2 Sampling strategies 76

Table 5.3 Factors used in $\bar{X}$ and R study 84

Table 5.4 Linearity data 90

Table 5.5 Example of measurement order 92

Table 5.6 Values for Z_O at a given confidence level 93

Table 5.7 Individual heights measured in inches 99

Table 5.8 Height data frequency diagram 99

Table 5.9 Short- and long-term sigma 103

Table 6.1 Data to analyze the effect of vitamin consumption 110

Table 6.2 Summary of sums of squares and degrees of freedom 112

Table 8.1 Control chart constants 132

Table 9.1 Factors for tolerance intervals values of K for two-sided
intervals 179

Table 10.1 R-DMAIC-SI strategy...188

Table 10.2 Techniques of facing a ball that a batsman should learn well 191

Table 10.3 Measurement datasheet 197

Table 10.4 Action plan..200

Table 10.5 Continuous improvement data 201

Foreword

1. If you believe in a product, don't give it up halfway through. Be on it. And you will succeed one day and the results will be good.
2. Have patience during difficult times. Don't lose your balance, and try to carry the team with you.
3. If it is a new business, plan for 50% more to standby so that you don't have to close the business or run away.

There is a lot of scope in manufacturing. The world's emerging economies can become strong in the long run only through a manufacturing base, not with a service base. A service base is only temporary. This will not create long-term employment.

Suresh has written four books—*Global Quality Management System*, *Lean for Operational Excellence*, *Six Sigma—Tactical Methodology*, and *Business Excellence—A Business Strategy*.

When I met Suresh and came to know about his operational excellence experience of more than two decades with multinational corporations (MNCs) such as Eaton Corporation and Fiat Global, a bell rang inside me and I made up my mind not only to pen the foreword, but also to leverage his Spanish language command to boost the performance of one of my South American Chilean units engaged in manufacturing wear-resistant products and material handling for the mining industry.

I knew Suresh well when I invited him to our Kolkata headquarters to spend one week at the Tega Head Office and the main plant at Joka in Kolkata, West Bengal, India. It was evident from the feedback report I received from my plant management team that this four-book series will clear the "cobwebs" and prepare any organization for the journey of continuous quality improvement.

These four books are a unique and comprehensive guide on how to understand and implement the Global Quality Management System (GQMS), Lean system, Six Sigma methodology, and business excellence strategy to achieve world-class business excellence. The author has succinctly summarized the business excellence concept and the body of knowledge of this book series by illustrating the business excellence pyramid with a foundation of the management system at the system level, the Lean system at the operational level, Six Sigma methodology at the tactical level, and business excellence at the strategy level.

Global Quality Management System starts by paying homage to leading quality "gurus." Having illustrated systems thinking as opposed to the command and control system, the author then stresses the fact that the command and control system can at the worst "influence people to behave in ways which dissatisfy the customer and/or suboptimize performance."

The main focus of any quality management system is on the process. The book stresses the importance of the process—its identification, definition, improvement, and control—using a turtle diagram and its extension to suppliers, inputs, process, outputs, and customers (SIPOC) diagrams. The processes discussed include, among others, main business processes such as human resources (HR), finance, project management, and importantly, the "process of improving the process."

Every documented GQMS has a focus on customer requirements and management system processes that lead to customer satisfaction. To this end, the author has included advanced processes to comply with ISO 9001, ISO/TS 16949, and AS 9100 standards, and elaborated on management improvement through extensive plan–do–check–act (PDCA) analysis and the problem-solving methodology involving the famous eight-discipline process (8D). The check and act phases are discussed extensively through audit processes and a process control plan audit (PCPA) process—as practiced by most automotive and MNCs.

The second book is about the Lean system. Part 1 explains why Lean implementation usually fails. It goes to show the approach for Lean transformation by highlighting the "cultural enablers" for employees (including management) and how management should align to the Lean transformation process. Part 2 is about Lean tools and how they can be deployed for continuous improvement. Part 3 is about Lean performance measures and how to assess Lean performance. Assessment of Lean system tools is a very interesting feature of this part and enables an organization to remain focused on the standardization of the Lean system and boost the organization's sustainability efforts.

The author has succinctly portrayed the main principles of the Lean system as follows:

1. Define customer requirements correctly and arrive at customer value so you are providing what the customer actually wants.
2. Identify the value stream for each product or service family and remove the non-value-added (wasted) steps for which the customer will not pay and that don't create value.
3. Make the value stream flow continuously to shorten throughput and delivery time aggressively.
4. Allow the customer to pull the product or service from your value streams as needed (rather than pushing products toward the customer on the basis of forecasts).
5. Never relent until you reach perfection, which is the delivery of pure value instantaneously with zero waste and zero defects.

The third book is about the unique way in which the so-called difficult concept and practice of Six Sigma methodology are depicted. It includes the collection of tools needed for all five phases—define, measure, analyze, improve, and control (DMAIC)—and proven best practices to identify which few process and input variables influence the process output measures. To begin with, the author describes the basic concepts of variation, spread of data, and sigma through basic statistical concepts. Before embarking on the five phases (DMAIC), the author clarifies what is needed for business performance measurement through the concepts of the balanced scorecard and important measuring units for quality performance. Notable measures discussed are defects per million opportunities (DPMO), rolled throughput yield, cost of poor quality (COPQ), business failure costs, cost–benefit analysis, return on assets (ROA), and last, a method of evaluating projects and investments known as net present value (NPV) or discounted cash flow (DCF).

The five phases of DMAIC form the bulk of the third book. The step-by-step approach taken by the author to explain the key concepts and tools required in each of these phases requires a special mention.

Define: This phase starts with the definition of the voice of the customer (VOC). Here the quality function deployment (QFD) tool is described in a simple and easy way to translate the customer's voice into the language of the engineer. The QFD is then utilized to define and document a business improvement project charter based on the customer and competitive intelligence data. Project tracking tools like the Gantt chart, critical path analysis (CPA), and the project evaluation and review technique (PERT) are explained in detail.

The CTQ flow-down is introduced to define customer satisfaction in four areas: quality, delivery, cost, and safety for internal and external customers.

Measure: The author has identified and discussed 16 different aspects of process characteristics. Having done this, he describes measurement system analysis (MSA) in great detail to ensure that the integrity of the measured data of important characteristics, the measuring equipment, and the human aspect of the measurement system are maintained within allowed repeatability and reproducibility (R&R) acceptance criteria.

Analyze: Here the root cause analysis methods for the problems encountered are discussed. The main techniques described include regression and correlation, analysis of variance (ANOVA), failure mode and effects analysis (FMEA), gap analysis, waste analysis, and Kaizen.

Improve: The process improvement methods discussed in this phase are prioritization through the cause-and-effect (C&F) matrix, Kaizen using Lean tools and Six Sigma, PDCA, and the theory of constraints.

Control: The key concepts and tools illustrated in the control phase are statistical process control (SPC), total productive maintenance (TPM) and overall equipment effectiveness (OEE), MSA, control plan, and visual factory. In order to sustain the improvements, the tools referred to are lessons learned, the training plan, standard operating procedures (SOPs), the work instruction, and ongoing performance assessment.

Design for Six Sigma (DFSS) methodology is a very useful and logical extension of the Six Sigma phases. The tools discussed in DFSS include define, measure, analyze, design, and verify (DMADV) and define, measure, analyze, design, optimize, and verify (DMADOV). Design for X (DFX) includes reliability analysis and design of tolerance limits. Special design tools described are Porter's five forces analysis and TRIZ (Russian for "theory of inventive problem solving").

No Six Sigma book can be called complete without a case study. To this end, the author has chosen an improvement project to improve batting in the game of cricket using the Lean Six Sigma approach.

The fourth book is about business excellence strategy. There are many models of business excellence practiced by many countries of the world.

At best, these models lay down business excellence assessment criteria, but the author feels that the main requirement of organizations intending to embark on business strategy is that they need a special body of knowledge with which the business excellence strategy can be implemented successfully. The inclusion of strategies for leadership, strategic planning, customer excellence, operational excellence, and functional excellence for HR and information technology (IT) will prove to be very useful for initiated management. There is a very effective chapter on the assessment of the business excellence strategy through the use of the balanced scorecard, employee surveys, achieving performance excellence, and cost-out.

Finally, as you put these four books of knowledge into practice, you will find that the roles of leaders and managers in your organization shift. It is not enough for the leaders to keep on doing what they have always done. It is not enough for them to merely support the work of others. Rather, leaders must lead the cultural transformation and change the mindsets of their associates by building on the principles behind all these excellent tools.

The author's account of these difficult and vast subjects is very praiseworthy and proof of his vast industrial experience of more than four decades of working with MNCs in Asia, Europe, and the Americas. This is an inspirational work that is easy to be learn and apply by the lay reader. I highly recommend this book to all students, teachers, executives, and organizations who want to learn and implement GQMS Lean Six Sigma systems and business excellence strategies.

Madan Mohanka
Chairman and Managing Director and Founder
TEGA Industries Limited
Kolkata, West Bengal, India

Preface

This book is about business excellence strategy. There are many models of business excellence practiced by many countries of the world. At best, these models lay down business excellence assessment criteria, but the author has felt that the main requirement of the organizations intending to embark on business excellence needs a special body of knowledge with which the business excellence strategy can be implemented successfully throughout the organization. The inclusion of strategies for leadership, strategic planning, customer excellence, operational excellence, and functional excellence for HR supported by a strong performance measurement, analysis, and knowledge management will prove to be very useful for the initiated management. The chapter on strategic planning promotes a virtuous cycle of trust and accountability resulting in employee empowerment and sound organizational capability.

Assessment of business excellence strategy through the use of the balanced score card, employee survey, achieving performance excellence, and cost out is part of the organizational infrastructure uniting people and motivating them to achieve planned organizational goals.

Throughout the book, time-tested principles and practices are applied to the business system and processes that deliver value to customers. The result is a holistic business excellence strategy.

Almost all business organizations are engaged in providing services or products to their customers. But when it comes to providing service to customers and presenting them an experience that will make them come back time and time again, only a small minority of organizations stand out from the crowd who apply the energy, commitment, and innovative thinking to get it right. There is an enormous difference between those who are truly focused on customer and those who simply pay lip service.

This book prepares the initiated person/organization for the journey of business excellence. The guiding principle is

> "An organization must constantly measure the effectiveness of its processes and strive to meet more difficult objectives to satisfy customers."

Taiichi Ohno
Toyota Production System

Acknowledgments

Acknowledging help and guidance in writing this four-book series is, to me, like churning the oceans of the world and putting all the blessings in a tea cup. I find it very daunting because during my more than 50 years of industry experience, I have been guided and helped by many persons, companies, and institutions with whose associations I have learned, practiced, and taught these subjects and achieved modest to excellent results.

After I decided to return to Ahmedabad from Texas, R.D. Patel, finance professor at the Indian Institute of Management (IIM) Ahmedabad, asked me to address their small and medium enterprises program as a guest speaker to talk about Lean Six Sigma. The feedback from the attendees was good, and Professor Patel took me to the Ahmedabad Management Association (AMA) to meet with the executive director of the AMA, K.K. Nair, who asked me to conduct the first-ever three-day AMA Lean Six Sigma seminar attended by industry representatives from Rajkot, Vadodara, Surat, and Ahmedabad. This led to another seminar at the AMA and an invitation by the human resources (HR) head of the Indian Space Research Organization (ISRO) (equivalent to U.S. NASA), J. Ravisankar, to address ISRO technicians and engineers on the subject of zero defects delivery of space systems, which was well received.

All of the above made K.K. Nair ask me to write a book on Lean Six Sigma for Indian engineers. My learning and experience as an operations excellence and engineering manager at Eaton Corporation (Eden Prairie, Minnesota) and Fiat Global (Burr Ridge Operations, Chicago) made me take a holistic view and include the Global Quality Management System at the bottom rung and business excellence at the top level. This has resulted in a four-book series.

I thank the following individuals for their contributions to my knowledge and all the help and guidance they offered to me in my career that resulted in creating this book series: C.S. Patel, former CEO of the Anand group of leading automobile companies manufacturing automotive components; the late D.N. Sarkar, chairman and managing director of Gestetner Limited; Samir Kagalwala, consultant for the design and manufacture of power magnetics; Stefan Lorincz, renowned electronics engineer and source developer for key electronic components worldwide at Phillips, Holland; Levy Katir, former Motorola vice president, who in 1994 put me in charge of quality and reliability of the newly developed electronic ballasts; G.P. Reddy, former director of quality at Universal Lighting Technologies; Inder Khatter, international quality management system lead auditor for DNV, Houston, Texas; Dev Raheja, international consultant and author of *Assurance Technologies: Principles and Practices*; Frank Kobyluch, global general manager at Klein Tools and former plant manager at Eaton Corporation; and Don Johnson, director of quality at Fiat Global, Case New Holland Division.

My special thanks and gratitude go to my colleagues and team members at the following companies, where I worked, learned, and developed and implemented many of the tools and techniques contained in this book series: Gestetner Limited (now Ricoh India); Energy Savings, Inc., Schaumburg, Illinois; United Lighting Technologies, Nashville, Tennessee; Eaton Hydraulics, Eden Prairie, Minnesota; and Fiat Global, Case New Holland, Burr Ridge, Chicago.

My abilities as an operations excellence manager in charge of providing quality products for compact fluorescent lamp (CFL) ballasts, hydraulic valves, pumps, hydraulic hoses, and fittings were honed, tested, and appreciated by customers such as GE CFL Lamps, Osram—Sylvania, John Deere, Case New Holland, Oshkosh Corporation, and several manufacturers of heavy-duty all-wheel-drive defense trucks: Caterpillar, GM Trucks, Ford Trucks, Volvo Trucks, Zamboni (ice resurfacer for the Olympic Games), and so forth.

I remain grateful to the following suppliers, who collaborated with me and my team in developing components and major assembly units requiring extremely high precision and pre- and posttreatments: Parker Hannifin, supplier of high-quality hydraulic seals and O-rings; Bosch, supplier of specialty hydraulic valves; Carraro Pune, supplier of a complete four-speed transmission unit for agricultural tractors; TGL-Carraro Pune, developer of precision gears and shafts for transmissions; Carraro, Quingdao, China, with whom we developed an entire rear-axle assembly for backhoe loaders; Graziano, Noida, where we developed a continuously variable transmission unit for a tractor for the first time for the U.S. market; GNA Group Punjab, supplier of forged and precision machined components for the tractor transmission assemblies; and Craftsman Automation Limited Coimbatore, who machined our large castings for transmission bodies and covers using heavy computer numerical control (CNC) machines and digital coordinate measuring machines (CMMs).

I have remained in touch with developing technology and professional knowledge through the American Society for Quality, whose membership I have held since 1993.

Illustrations and the design of charts and figures in this book series were done by Sanjay Trivedi and Minal Mehta.

Making It Big in Manufacturing Products and Providing Service

It is a general belief that successful people in every field are blessed with talent or are just lucky. But the fact is that successful people work hard, work long, and work smart.

Marissa Ann Mayer, the current president and CEO of Yahoo, used to work 130 hours per week while working with Google. India-born Indra Krishnamurthy Nooyi, the chairman and CEO of PepsiCo, worked midnight to 5:00 a.m. as a receptionist to earn money so that she could complete her master's degree at Yale University. In 1958, Qimat Rai Gupta left his education midway through and founded electric trading operations in the electric wholesale market of Old Delhi, India. With an investment of Rs 10,000 (US$150), he started Havells. Today Havells is a billion dollar company. In his own words, "Overnight success means 25 years of hard work, devotion and dedication."

The story of the founder and CEO of Tega Industries, based in Kolkata, India, Madan Mohanka, is unique. When he went into business, he had the right combination—hailing from a business family, having an engineering degree, earning his MBA from the Indian Institute of Management (IIM) Ahmedabad, and having a foreign collaboration as a joint partner. Yet this combination failed miserably. He witnessed the imminent closure of his company in 1979, but like the epic hero Odysseus, he never lost focus. He kept at it. Some three decades later, it was Madan's die-hard optimism that saw Tega Industries become the second largest player in the world in rubber mill lining products for the mining industry.

In her book *Stay Hungry, Stay Foolish*, Rashmi Bansal (IIM Ahmedabad graduate) depicted Madan Mohanka's hard-won story very aptly. She said Madan faced all the hurdles and challenges of starting from scratch, but then Madan had what you call an obsession. Over the last three decades, Madan built a strong foundation combining three technologies: mechanical engineering, rubber (polymer) technology, and mineral processing and grinding. In recent years, Tega has accepted challenges, grabbed overseas marketing opportunities, and maintained consistent growth, keeping an eye on the margins.

Tega's presence in 19 international locations has enabled it to increase a turnover of Rs 23 crore in 2009 to Rs 681 crore in 2014.

According to Mehul Mohanka, his U.S.-trained MBA son, the stage is now set for organic and inorganic growth—organically building up larger capabilities and inorganically looking for acquisitions for successful integration with Tega's culture, values, and philosophy.

1

Introduction to Six Sigma: A Tactical Strategy

ORIGIN OF SIX SIGMA AT MOTOROLA

The evolution of Six Sigma began in the late 1970s at a Motorola plant in Chicago. The revealing moment came when a Japanese firm took over a Motorola factory that manufactured television sets in the United States. Under Japanese management, the factory was soon producing TV sets with 1/20 the number of defects they had produced under Motorola management. Inspired by this, Bob Galvin, Motorola's CEO in 1981, challenged his company to achieve a 10-fold improvement in performance over a five-year period. To achieve this, they needed to halve the defects each year. This initiative and subsequent hard work by Motorola engineers resulted in a worldwide tactical strategy for business excellence called Six Sigma.

The credit for coining the term *Six Sigma* goes to Bill Smith, a senior engineer at Motorola. In 1984, he discovered the correlation between how well the product did in its field life and how much rework and repairs were required during the manufacturing process. He found that products that were built with fewer nonconformities or those that had less variation in their specifications during manufacture were the ones that performed the best after delivery to the customer.

Bill Smith and Mikel Harry developed a four-stage problem-solving approach—measure, analyze, improve, and control (MAIC)—to find and reduce the variations, thus eliminating defects altogether.

Later, when the define stage was added, the DMAIC discipline became the road map for achieving Six Sigma quality.

UNDERSTANDING THE VARIATION, SPREAD, AND SIGMA

The goal of most processes in the manufacturing and service industries is to produce products or services that have little to no variation. Variation is defined as no two items or services being exactly the same. Variation also can be defined as the extent to which or the range within which something varies.

Variation impacts performance and cost. It makes products and processes difficult, unpredictable, untrustworthy, and of poor quality. Good quality is strongly tied to reliability, trustworthiness, and no unpleasant surprises. In other words, bad quality results from too much variation, and good quality results from little variation.

It is important to reduce variation because of the economic loss resulting from customer dissatisfaction and poor quality.

Variation is defined as an inevitable change in the output or result of a system (process) because all systems vary over time. Two major types of variations are (1) common, which is inherent in a system, and (2) special, which is caused by changes in the circumstances or environment (Figures 1.1 and 1.2).

Defects, delays, and deviations (fluctuations) in a system or a process are caused by variation.

Common causes to process variation are tool wear, machine vibration, and changes in work-holding devices. Changes in material composition and hardness are also sources of variation.

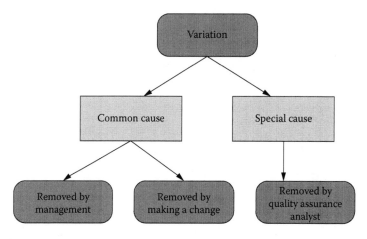

FIGURE 1.1
Types of variations and their methods of removal.

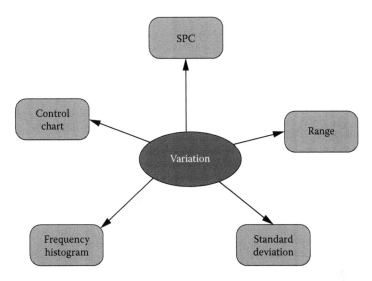

FIGURE 1.2
Various means of measuring variation.

Special causes to process variation are overadjusting the machine by the operator, making an error during the inspection activity, changing the machine settings, or failing to properly align the part before machining.

Examples of environmental factors affecting variation include heat, light, radiation, and humidity.

With statistical methods, several types of variations can be tracked. These include

Within-piece variation: This is the variation within a single item or surface.

Piece-to-piece variation: This is the variation that occurs among pieces that are produced at about the same time.

Time-to-time variation: This is the variation in the product produced at different times of the day.

Product or process spread: This is a long-term variation in a product or process.

Lot-to-lot variation: This is the variation that occurs among lots that are produced at about the same or different times.

Interaction variations: These are variations found in the interaction between man and machine, inspector and the measurement device, or materials used and the environment.

As the spread of variation becomes wider from the limits specified by the customer, or the mean or average value of the data, more defects with less yield (defect-free product) are created.

Six Sigma is the best available method to reduce process variability. The steps are

- Identify the needs of the customer.
- Translate these needs into the process expert's language through quality function deployment (QFD).
- Make improvements through the DMAIC process.
- Hold the gains through statistical process control (SPC).
- Provide customer satisfaction.

_____

DMAIC PROCESS

The concepts of independent and dependent variables can be explained as follows. An independent variable is a factor or event that causes or affects another related factor or event called a dependent variable. For example, exercise is an independent variable (X) because it influences another variable called body weight, a dependent variable (Y). In an experiment, it is the independent variable (X) that is allowed to change in a systematic manner, and its effect on the behavior of a dependent (controlled) variable (Y) is studied. Y is also called an output variable. So output Y is dependent on X:

$$Y = f(x)$$

The process outcome (Y) is a result of the process inputs or process drivers (X) within the process. The aim of DMAIC is to identify which few input process variables influence the process outputs (Figures 1.3 and 1.4).

Define: Understand the project output Y and how to measure it.
Measure: Priority-wise, determine potential Xs and measure Xs and Y.
Analyze: Determine X–Y relationships and, after verification, quantify important Xs.
Improve: Devise solutions to optimize Xs to improve Y.
Control: Control and monitor important Xs and the output Y over time.

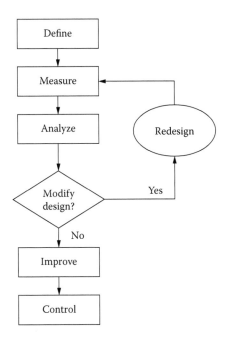

FIGURE 1.3
DMAIC process.

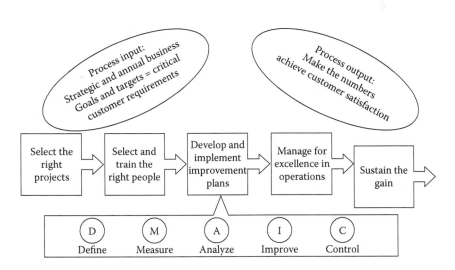

FIGURE 1.4
Generic Six Sigma process.

2

What Is Six Sigma?

For the benefit of all, let us review some basic statistics to understand the concept of sigma.

BASIC STATISTICS

Types of Statistics

Dr. Deming discussed the importance of the difference between enumerative and analytic studies (also known as statistics). The basic difference is this:

> Descriptive (enumerative) statistics describe data using math and graphs, and focus on the current situation, such as measuring a sample and then estimating the population's characteristics.
>
> Inferential (analytic) statistics use sample data to predict or estimate what a population will do in the future, such as measuring periodic samples from a process that is continually manufacturing to predict the results of the next batch.

It may be helpful to consider these two examples:

1. A tailor takes a measurement (waist, chest, inseam, etc.) from a customer who purchases a new suit.
2. A doctor takes a measurement (temperature, blood pressure, heartbeat, etc.) from a patient who feels ill.

In the first instance, the tailor is taking a measurement to obtain current quantifiable information. The tailor is using a descriptive approach.

In the second example, the doctor is taking a measurement to obtain a causal explanation for some observed phenomenon. The doctor is taking an analytic approach.

We use descriptive statistics to describe data, usually sample data, with math or graphics to define elements such as

- Central tendency of the data: can be measured by median, mean, and mode
- Data variation: can be measured by range of the data and variance
- Graphs of the data: histograms, box plots, and so forth

Analytical Statistics

Analytical statistics involves the evaluation of ratio (or measured) data. Analytical statistics are usually performed to estimate the population parameters (estimation), to determine the difference between two populations (hypothesis testing), to determine the differences among a number of populations (analysis of variance), or to evaluate the degree of relationship between two or more variables (correlation and regression).

Analytical statistics are usually performed using the scientific process:

1. Make a hypothesis of what we expect to find.
2. Collect data.
3. Analyze the data.
4. Draw a conclusion about the validity of the hypothesis.

So analytical statistics describes what the population should be in order to have given rise to the sample that was obtained.

For example, if a sample of four taken from a box of bubble gum is found to have three orange pieces and one red, we can conclude that the box contains 75% orange bubble gum. Although this may or may not be correct, a conclusion is drawn. A larger sample would provide a better estimate.

DEFINITION OF SAMPLE STATISTIC, POPULATION, AND POPULATION PARAMETER

A *statistic* is a quantity derived from a sample of data that assists in forming an opinion of a specified parameter of a target population. A sample is frequently used because data on every member of a population are often impossible or too costly to collect.

A *population* is an entire group of objects that have been made, or will be made, containing a characteristic of interest.

A *population parameter* is a constant or coefficient that describes some characteristic of a target population. An example of a population parameter is the mean or variance.

Frequently used symbols are:

Sample	Population
n = Sample size	N = Population size
$\bar{x}$ (X-bar) = Sample mean	μ = Population mean
S = Sample standard deviation	σ = Population standard deviation
S^2 = Sample variance (square of sample standard deviation)	σ^2 = Population variance (square of population standard deviation)

CENTRAL LIMIT THEOREM AND STANDARD NORMAL DISTRIBUTION

The *central limit theorem* is the theoretical foundation for many statistical procedures. The theorem states that a plot of the sampled mean values from a population tends to be normally distributed (Figure 2.1).

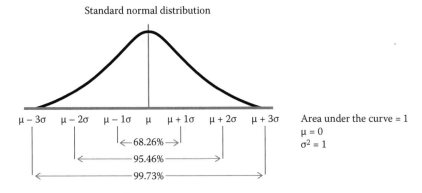

Standard normal distribution

$\mu - 3\sigma$	$\mu - 2\sigma$	$\mu - 1\sigma$	μ	$\mu + 1\sigma$	$\mu + 2\sigma$	$\mu + 3\sigma$

←68.26%→
←95.46%→
←99.73%→

Area under the curve = 1
$\mu = 0$
$\sigma^2 = 1$

FIGURE 2.1
Normal distribution.

Key points of the central limit theorem and Six Sigma include:

- Using ±3-sigma control limits, the central limit theorem is the basis of the prediction that if the process has not changed, a sample mean falls outside the control limits on an average of only 0.27% of the time. Or the process yield at the 3-sigma level is 99.73% and the defective product is 0.27%.
- Sixty-eight percent of values are within 1 standard deviation of the mean.
- Ninety-five percent of values are within 2 standard deviations of the mean.
- A total of 99.7% of values are within 3 standard deviations of the mean.

A *standard normal distribution* has the following characteristics:

- Most points on the curve tend to be near the average.
- The curve's shape tends to be bell shaped, and the sides tend to be symmetrical.

The normal distribution has

- mean = median = mode.
- Symmetry about the center.
- Fifty percent of values less than the mean and 50% greater than the mean.
- The theorem allows the use of smaller sample averages to evaluate any process because distributions of sample means tend to form a normal distribution.
- The theorem appears when the process is in control (predictable).
- The theorem leaves variations from common causes to chance.
- The theorem identifies and removes variations from special causes.

As can be seen in the bell-shaped curve in Figure 2.2, the distance between the mean line and the inflection point (where the bell curve changes the direction from going out to coming in) is called 1 sigma (σ). In other words, 1 sigma measures the data spread from the mean to the inflection point.

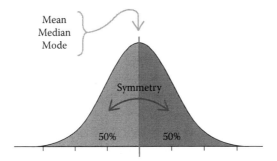

FIGURE 2.2
Normal distribution is symmetrical.

The bell-shaped curve, also called the normal curve, shows specification values concentrated in the center and decreasing specification values on either side. This means that the data have less tendency to produce unusually extreme values, compared to some other distributions.

The bell-shaped curve is symmetrical. This tells you that the probability of deviations from the mean are comparably similar in either direction. The curve represents the distribution of any of the following of a set of data:

- Values
- Frequencies
- Probabilities

It slopes downward from a top point in the middle (mean value) or the maximum probability point (ideally 50%). Data that reflect the collective result of large numbers of unrelated events tend to result in bell curve distributions, for example, a dimension of a machined part on a computer numerical control (CNC) machine, service waiting time at McDonalds, or grades of students in a class.

CENTRAL TENDENCIES

Central tendency is a measure that characterizes the central value of a collection of data that tends to cluster somewhere between the high and

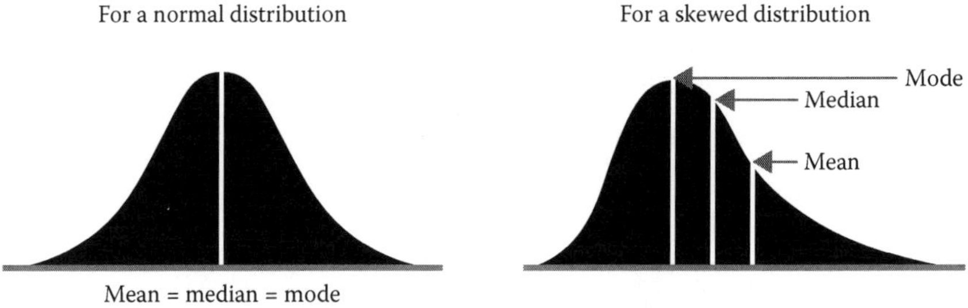

For a normal distribution   For a skewed distribution

Mean = median = mode

FIGURE 2.3
Mean, median, and mode in normal and skewed distributions.

low values in the data. Central tendency refers to a variety of key measurements, such as mean (the most common), median, and mode (Figure 2.3).

Mean
- Gives the distribution's arithmetic average (center)
- Provides a reference point for relating all other data points
- Is typically used with normal data

Median
- The distribution's center point (middle value)
- Equal number of data points occur on either side of the median
- Useful when the data set has extreme high or low values
- Typically used with nonnormal data

Mode
- Represents the value with the highest frequency of occurrence (the most often repeated value)
- Typically used with nonnormal data

MEASURE OF DISPERSION

In this section, we discuss measures of dispersion (Figure 2.4).
All at the same time, Six Sigma is

- A metric
- A methodology
- A management system

Measure of dispersion	Definition	Formula
Range (R)	The difference between the largest and smallest values in a data set.	$A - B = R$ A = largest value in data set B = smallest value in data set
Variance (σ^2, S^2)	The sum of the squared deviations from the mean, divided by the sample size or degree of freedom. Also, the standard deviation squared.	For a population: $\sigma^2 = \dfrac{\Sigma(x - \mu)^2}{N}$ For a sample: $S^2 = \dfrac{\Sigma(x - \bar{x})^2}{n - 1}$
Standard deviation (σ, S)	The computed measure of variability indicating the spread of the data set around team. Also, the square root of the variance.	For a population: $\sigma = \sqrt{\dfrac{\Sigma(x - \mu)^2}{N}}$ For a sample: $S = \sqrt{\dfrac{\Sigma(x - \bar{x})^2}{n - 1}}$

FIGURE 2.4
Formulas for measures of dispersion.

Six Sigma as a Metric

Six Sigma (6σ) is a measure of quality that is very close to perfection. The statistical meaning of Six Sigma is having 6 standard deviations spread between the mean and the nearest specification limit. A Six Sigma process is virtually defect-free, with only 3.4 defects in a million opportunities. The Six Sigma process is 99.9997% defect-free.

A defect is defined as anything outside of customer specifications.

An opportunity is a chance for a defect to occur. We can detect it, correct it, and prevent it from occurring again. Defects per million opportunities is termed DPMO.

Statistical Interpretation of Six Sigma

For the curve shown in Figure 2.5, the average or mean $\mu = 0$ and the standard deviation $\sigma = 1$. The upper (USL) and lower (LSL) specification limits are at a distance of 6 sigma from the mean. Because of the properties of the normal distribution, values lying that far away from the mean are extremely unlikely. Even if the mean were to move right or left by 1.5 sigma at some point in the future (1.5-sigma shift), there is still a good safety cushion. This is why Six Sigma aims to have processes where the mean is at least 6 sigma away from the nearest specification limit.

Any normal distribution can be converted to a standard normal distribution.

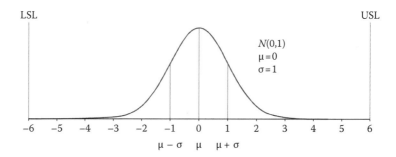

FIGURE 2.5
Standard normal distribution.

Here is the formula for *z*-score:

$$z = \frac{x - \mu}{\sigma}$$

where *z* is the *z*-score (standard score), *x* is the value to be standardized, μ is the mean, and σ is the standard deviation.

Why Is the Normal Distribution Useful?

Many things are actually normally distributed or very close to it. For example, height, weight, and intelligence of a population are approximately normally distributed. Dimensions of manufactured parts also often have a normal distribution.

The normal distribution is easy to work with mathematically. In many practical cases, the methods developed using normal theory work quite well even when the distribution is not normal.

There is a very strong connection between the size of a sample *N* and the extent to which a sampling distribution approaches the normal form. Many sampling distributions based on large *N* can be approximated by the normal distribution even though the population distribution itself is definitely not normal.

The term *sigma* is a letter taken from the Greek alphabet, equivalent to the English S. It is used to designate the distribution or spread on either side of the mean (average) of any parameter of a product, process, or procedure.

Sigma (σ) is used by statisticians to show the variation in a process.

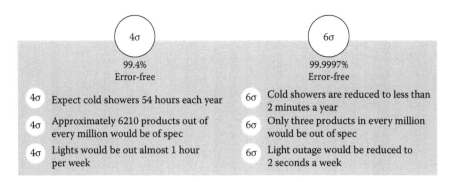

FIGURE 2.6
Significance of sigma level in defect assessment.

For example, let us bring this concept to life in our own home. The difference between 4 sigma and 6 sigma is illustrated below (Figure 2.6):

4 sigma means 100% – 99.379% = 0.6210% defectives

or

0.6210 defectives out of 100

or

0.6210 × 10,000 = 6210 defectives out of a million

or

6210 parts per million (ppm)

Note: 1% = 10,000 ppm.

Parts per Million and Sigma Level

Generally, most companies operate in the range of 3–4 sigma.

Notice in Figure 2.7 that as the sigma level increases, the parts per million rate decreases.

Here the data show that, on average, flight delays are 6000 ppm, or 0.6%, while airline fatalities are 1 ppm, or 0.0001%.

FIGURE 2.7

Relation between parts per million and sigma level. (Adapted from Breyfogle III, F.W., *Implementing Six Sigma: Smarter Solutions Using Statistical Methods*, 2nd ed., John Wiley & Sons, New York, 2003.)

TABLE 2.1

Sigma Level and DPMO

Sigma	DPMO
6.0	3.4
5.5	30
5.0	230
4.5	1350
4.0	6210
3.5	22,700
3.0	66,800
2.5	158,000
2.0	308,000
1.5	500,000
1.0	690,000
0.5	840,000

Table 2.1 shows the relationship between the sigma level and the defective parts per million.

Six Sigma as a Management System

Six Sigma is a performance management system for executing business strategy. It aligns improvement efforts to business strategy and business goal metrics. Six Sigma leverages meaningful metrics to monitor success.

Six Sigma as a Methodology

Six Sigma is a disciplined and data-driven methodology to improve, design, and manage processes, by focusing on improving business performance and meeting customer requirements.

We have seen that Lean helps a company cut the time it takes to meet customer demands and eliminate all kinds of delays and wastes that it faces during the process of providing the product or service to the customer.

Six Sigma is the "best of the best"—meaning a Six Sigma strategy takes the best of the quality principles and techniques used over many years and applies them appropriately to an organization. Six Sigma helps in finding out and fixing the mistakes, errors, defects, and deviations (variations) involved in every aspect of delivering customer wants. By taking the best principles and techniques from each of these programs, organizations maximize their productivity, profitability, growth, and improvements in measurable ways.

3

Business Performance Measures

ORGANIZATIONAL PROCESSES AND THEIR IMPACT ON THE ORGANIZATION

The Six Sigma methodology recognizes that there are many sources for inputs, outputs, and feedback for an organization. Each output may have its own process dependent on the input from other processes. All inputs and outputs of a process should be measurable so that quality can be controlled.

See Figure 3.1 to understand the organizational processes and their interaction in terms of their inputs, outputs, product, and information flow.

It is important to know that improvements in one area may create errors in another. So the understanding of interaction and interdependence of processes is critical to the success of a Six Sigma project.

DEFINING OWNERS AND STAKEHOLDERS

Top management participation is extremely important to the success of Six Sigma projects.

Quality improvement projects cannot happen without the appropriate decision makers taking ownership of the project. "Buying in" to the change by the decision makers and leaders is the first resource requirement.

In addition to these, other resources must be available to give an appropriate level of input and make decisions. Often, the resources that are needed are the busiest and will not find time to participate in the project unless its level of importance is appropriately elevated. The following resources as stakeholders should be identified early on and informed of their role in the project as soon as practical. A resource must commit to a certain time as required by the project.

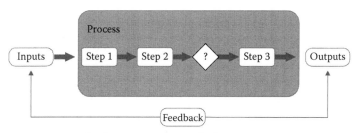

Conduct measurements at both points to gage
the efficiency and effectiveness of the process.

FIGURE 3.1
Organizational process with feedback loop.

The stakeholders are:

Customers
Employees (internal customers)
Suppliers
Investors
Government
Community

BALANCED SCORECARD

A balanced scorecard (BSC) is a widely accepted approach to establishing
an organizational strategy (Figure 3.2). It is a strategic measurement and
management system that translates an organization's strategy into four
perspectives (Table 3.1).

Figure 3.3 shows an example of a typical balanced scorecard.

KEY PERFORMANCE INDICATORS

Every organization wishing to embark on a Six Sigma journey must set-
tle on a few key performance indicators (KPIs) to use for planning and

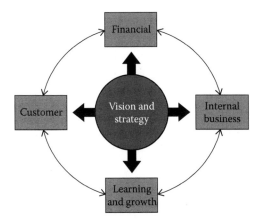

FIGURE 3.2
Balanced scorecard diagram.

TABLE 3.1

Four Perspectives to Determine Organizational Strategy

Perspective	Strategy
Financial	To achieve financial success, ask, "How should we appear to our shareholders?" They should say, "This organization is one of my best investments."
Customer	To achieve our vision, ask, "How should we appear to our customers and employees?" Customers should say, "We want to do more business with this organization." Employees should say, "I am proud to be part of this team."
Internal business process	To achieve shareholder and customer satisfaction, ask, "What business processes must we excel at?"
Learning and growth	To achieve our vision, ask, "How will we sustain our ability to change and improve?"

guidance, as well as use other metrics called key support indicators (KSIs) in the role of supporting the KPIs.

There are numerous performance indicators used in Six Sigma projects. The metrics for these performance indicators are also known as critical-to-quality (CTQ) measures. The CTQ flow-down or the performance indicator flow-down of a typical organization is further explained in Chapter 4 on the define phase.

Scorecard Metric	2004 Actual	2005 Actual	2006 Actual	2007 Goal	Jan	etc.....
1 Total Sales ($,000)	$48,081	$53,133	$62,925	$0	$5,368	
2 Past Due Shipments ($,000 past due)	$665	$702	$651	$0	$1,020	
3 Mix (% Distribution Sales)	26%	26%	33%	0%	25%	
Customer Satisfaction/Quality						
4 Customer Returns (DPPM)	457	4,914	209	0	393	
5 On Time Delivery (% to MADD)	83.1%	79.2%	78.4%	0.0%	77.8%	
6 Final Test Failure Rate (DPPM)	45,870	27,288	5,590	0	16,421	
7 Scrap (as % of Sales)	0.29%	0.30%	0.49%	0.00%	0.99%	
Innovation						
8 Cost Improvement ($,000)	$700	$743	$1,858	$0	$189	
9 Lean -8 Score	3.3	3.6	3.3	0.0	3.3	
Cycle Time						
10 Total Inventory - DOH	39.4	45.6	56.3	0.0	58.3	
Suppliers						
11 Supplier Quality (DPPM)	4,980	2,557	462	0	1,152	
12 Supplier Delivery (% On Time)	90.6%	84.7%	80.4%	0.0%	95.4%	
Employee Satisfaction						
13 Safety (OSHA Recordable Incident Rate)	0.00	0.00	0.21	0.00	2.64	
14 EPWP Score			3.0	0.0	2.8	
15 Flex Training Delivered (YTD %)			100%	0%	0%	
16 MESH Assessment Score			100%	0%	89%	
Business Excellence						
17 Cost of Nonconformance (% of Sales)			6.5%	0.0%	4.0%	
18 Incremental/Decremental ($,000)	($1,505)	($3,010)	($433)	$0	-$88	
19 CFROGC	21.50%	3.90%	5.80%	0.00%	15.20%	
20 Manufacturing Profit (% of Sales)	13.40%	13.30%	14.32%	0.00%	14.38%	
Location-Specific Metrics						
21 Rolled Throughout Yield (RTY %)			98.9%	0.0%	99.5%	
22 Prototype Delivery (% On Time)			83%	0%	88%	
				0.0		

FIGURE 3.3

Example of a typical balanced scorecard. CFROGC, cash flow return on average gross capital; DOH, days on hand; DPPM, defective parts per million; MADD, mutually agreed delivery date; OSHA, Occupational Safety and Health Administration; RTY, rolled throughput yield; YTD, year to date.

IMPORTANT MEASURES OF QUALITY PERFORMANCE

Defects per Unit

As an example, a process has produced 40,000 ball pens. According to the process engineer, three types of defects can occur.

The number of occurrences of each defect type are

Blurred printing: 36
Wrong dimensions: 118

Rolled ends: 11
Total number of defects: 165

$$\text{Defects per unit (DPU)} = \frac{\text{Total number of defects}}{\text{Total number of units produced}} = 165/40,000$$

$$= 0.004125$$

Defects per Million Opportunities

Each opportunity to make a defect must be counted. To calculate the number of opportunities, it is necessary to find the number of ways each defect can occur in each item.

In the pen product, blurred printing occurs in only one way (the ink not flowing properly), so in the batch there are 40,000 opportunities for this defect to occur.

There are three independent places where dimensions are checked, so in the batch there are 3 × 40,000 = 120,000 opportunities for dimensional defects.

Rolled ends can occur at the top or bottom of the ball pen, so there are 2 × 40,000 = 80,000 opportunities for this defect to occur.

Thus, the total number of opportunities for defects is 40,000 + 120,000 + 80,000 = 240,000.

$$\text{DPMO} = \frac{(\text{Number of defects}) * (1,000,000)}{\text{Total number of opportunities}} = 165,000,000/240,000 = 687.5$$

Throughput Yield

The throughput yield is the exponential function $= e^{-\text{DPU}}$. In the pen example, the computation for the throughput yield is $e^{-0.004125} = 0.996$. In other words, there is a probability of 99.6% that the product will be as per specifications.

Rolled Throughput Yield

The rolled throughput yield (RTY) applies to the yield from a series of processes and is found by multiplying the individual process yields. If a product goes through four processes whose yields are 0.994, 0.987, 0.951, and 0.990, then

$$RTY = (0.994)(0.987)(0.951)(0.99) = 0.924$$

Parts per Million

Parts per million (ppm) is defined simply as

$$ppm = DPU \times 1,000,000$$

In this example, the computation for the parts per million is

$$ppm = (0.004125)(1,000,000)$$

$$ppm = 4125$$

Parts per million is also used to refer to contaminants, as seen in the following example.

Suppose 0.23 g of sand particles is found in 25 kg of product:

$$PM = \left(\frac{0.23}{25,000} \right) \times 1,000,000$$

Table 3.2 shows the relation among DPMO, sigma, and process capability index (Cpk).

Table 3.3 shows the relations between sigma level, process yields, and short- and long-term capabilities.

Cost of Poor Quality

The cost of poor quality (COPQ), also known as business failure cost, calculates the financial impact of defects (Figure 3.4). This is helpful in prioritizing projects in terms of their impact on the enterprise. COPQ should include the expenses involved in rework, warranty, late deliveries, customer dissatisfaction, and so on.

TABLE 3.2

Sigma Conversion Table Showing Relations among DPMO, Sigma, and Cpk

Defects per Million Opportunities	Sigma Level (with 1.5-Sigma Shift)[a]	Cpk (Sigma Level/3) (with 1.5-Sigma Shift)[a]
933,200	0.000	0.000
915,450	0.125	0.042
894,400	0.250	0.083
869,700	0.375	0.125
841,300	0.500	0.167
809,200	0.625	0.208
773,400	0.750	0.250
734,050	0.875	0.292
691,500	1.000	0.333
645,650	1.125	0.375
598,700	1.250	0.417
549,750	1.375	0.458
500,000	1.500	0.500
450,250	1.625	0.542
401,300	1.750	0.583
354,350	1.875	0.625
308,500	2.000	0.667
265,950	2.125	0.708
226,600	2.250	0.750
190,800	2.375	0.792
158,700	2.500	0.833
130,300	2.625	0.875
105,600	2.750	0.917
84,550	2.875	0.958
66,800	3.000	1.000
52,100	3.125	1.042
40,100	3.250	1.083
30,400	3.375	1.125
22,700	3.500	1.167
16,800	3.625	1.208
12,200	3.750	1.250
8800	3.875	1.292
6200	4.000	1.333
4350	4.125	1.375
3000	4.250	1.417
2050	4.375	1.458
1300	4.500	1.500

(Continued)

TABLE 3.2 (CONTINUED)

Sigma Conversion Table Showing Relations among DPMO, Sigma, and Cpk

Defects per Million Opportunities	Sigma Level (with 1.5-Sigma Shift)[a]	Cpk (Sigma Level/3) (with 1.5-Sigma Shift)[a]
900	4.625	1.542
600	4.750	1.583
400	4.875	1.625
230	5.000	1.667
180	5.125	1.708
130	5.250	1.750
80	5.375	1.792
30	5.500	1.833
23.4	5.625	1.875
16.7	5.750	1.917
10.1	5.875	1.958
3.4	6.00	2.000

Note: Long-term Cpk*3 = Sigma level.

[a] The table assumes a 1.5-sigma shift because processes tend to exhibit instability of that magnitude over time. In other words, although statistical tables indicate that 3.4 DPMO is achieved when 4.5 process standard deviations (sigma) are between the mean and the closest specification limit, the target is raised to 6.0 (4.5 + 1.5) standard deviations to accommodate adverse process shifts over time and still produce only 3.4 DPMO.

TABLE 3.3

Sigma, DPMO, Yield, and Short- and Long-Term Cpk Comparison

Sigma Level	DPMO	Percent Defective	Percentage Yield	Short-Term Cpk	Long-Term Cpk
1	691,462	69%	31%	0.33	−0.17
2	308,538	31%	69%	0.67	0.17
3	66,807	6.70%	93.30%	1	0.5
4	6210	0.62%	99.38%	1.33	0.83
5	233	0.02%	99.98%	1.67	1.17
6	3.4	0.00%	100.00%	2	1.5
7	0.019	0.00%	100.00%	2.33	1.83

Note: Long-term Cpk 1.5*3 = 4.5 sigma, which becomes 6 sigma when 1.5 is added to it.

COPQ calculates the financial impact of defects.
This is helpful in prioritizing projects in terms of their impact on the enterprise.

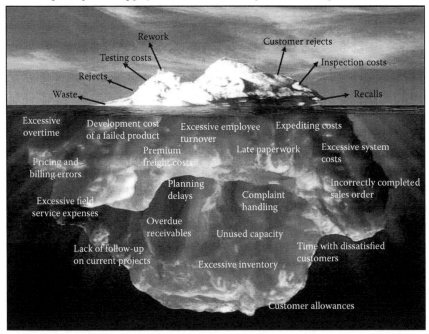

COPQ consists of 15%–25% of the total cost.

FIGURE 3.4
Cost of poor quality.

W. Edwards Deming is frequently quoted as saying that the most important numbers are unknown and unknowable. Who can state, for instance, the value of a satisfied customer who will, in Deming's words, "tell all his friends and some of his enemies?" Loyal customers are to be sought and cherished—he very life of the organization depends on them. To estimate the financial impact of a loyal customer, one would need to consider all the costs required to locate and recruit a new customer, including marketing and sales, and the time and effort to help make this new customer into a loyal customer.

Business Failure Cost Measurements in Manufacturing and Finance

Now let us look at these costs in a typical manufacturing business as a percentage of the revenue (Figure 3.5). Figure 3.6 shows that 25% of the total sales revenue is eaten away by the business failure costs. The right-hand column shows the hypothetical breakup of the business failure costs. For example, if the total sales revenue is 100, the business failure costs will amount to 25.

Manufacturing

Left		Right
Product service Liability	External failure	Warranty Policy adjustments
Scrap Rework Scrap and rework related to vendor Excess labor and materials	Internal failure	Failure investigation and correction Product and process redesign Downtime and delays
Test and inspection of purchased materials Measuring services Field testing Approval expenditures	Appraisal	Process or product evaluation and report Quality information equipment expense Quality audits

Quality and reliability
Analysis and planning
Quality training and manpower
  development
	Prevention	Specification design and
		  development of quality
		  information equipment

Finance

Left		Right
Lost business Write-offs Customer attrition Customer complaints Fraud	External failure	Legal or regularity failure Collections Resolving customer disputes Statement errors Net credit loss
Computer downtime Nonperforming assests Defective purchased materials and services	Internal failure	Data entry errors Hedging errors Rework Employee turnover

FIGURE 3.5
Business failure costs.

Out of 25, external failure costs will be 25% of 25, or 6.25% of the total revenue. In the same way, internal failure costs will amount to 50% of 25, or 12.5%; similarly the appraisal cost is 5% and the prevention cost 1.25% of the total revenue.

Now let us consider this COPQ as an improvement driver. After Kaizen exercises, assume that we can achieve an estimated 50% reduction in COPQ, or from 25% to 12.5% of the total revenue, as shown in Figure 3.7.

Figure 3.7 will now become the current state, and future states can be planned using Kaizen and the plan–do–check–act (PDCA) approach to reduce the COPQ further as planned organizational goals. Similarly, finance operation failure costs can be analyzed using Kaizen and the

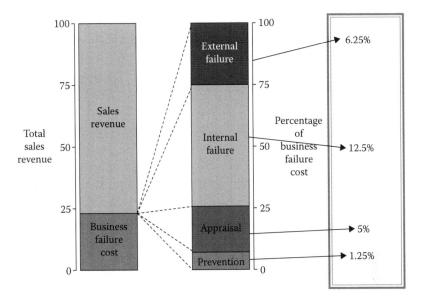

FIGURE 3.6
Business failure costs analyzed.

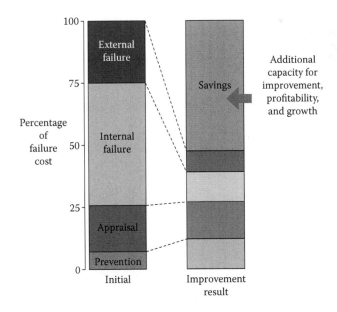

FIGURE 3.7
Business cost analysis as an improvement driver.

PDCA approach to reduce the cost of poor quality even more through planned organizational goals.

Financial Measures

Project Financial Benefits

When determining whether a Six Sigma project has been successful, the bottom line is often the factor that matters the most. The key question is, how much value does the organization realize as a direct result of this project's success?

Financial Benefits Defined

Just as there are costs associated with poor quality (see COPQ), there are also financial benefits of implementing changes that result from successful Six Sigma projects.

These financial benefits include

- Additional revenue from increased sales
- Cost avoidance or mitigation
- Faster return on investments
- Lower production costs
- Lower costs associated with customer service
- Increased cash flow
- Enhanced profitability of existing services or products
- Increased revenue of existing sources
- Increased value in organization stock or perceived value

The financial factors discussed here are

- Cost–benefit analysis
- Return on assets (ROA)
- Return on investment (ROI)
- Net present value (NPV)

Cost–Benefit Analysis

To perform a cost–benefit analysis for the project, complete these steps:

Step 1: Identify the project benefits. Quantify the expected financial benefits assuming the project obtains the expected result or reaches its goal.

Express the benefits in financial amounts with specific time limits.

For example, the project should realize an increase in sales of Rs 50 lacs per year beginning in year 2 and lasting for three years.

Step 2: Identify the project costs. Try to limit the project costs to the budgeted cost of the project.

Step 3: Calculate whether benefits exceed costs.

Step 4: Determine whether the project should be implemented. Even if the benefits do not exceed the costs, senior management may elect to complete the project. For example, improving the customer satisfaction rating of checking and savings account customers of the bank is a benefit that may not exceed the cost of a project, but could impact future consumer lending relationships.

If this is the case, determine what could be done to enhance the benefits and minimize the project costs.

A cost–benefit analysis may also be initiated after a project has been completed to determine whether the project should have been undertaken in the first place.

Cost–Benefit Analysis: Results

The results of a cost–benefit analysis may be any of the following:

- The overall benefits and costs are compared to determine whether benefits exceed costs on a straight money spent to money gained basis.
- The benefits and costs are used within a ratio such as ROA or ROI. On a relative basis, these ratios can express the expected return for use of the assets or the investment in the project.
- The information may also be used in an NPV equation if there is an issue with benefits and costs being spread out over a longer period of time.

Cost–Benefit Analysis: Hard Cash versus Soft Cash

Hard cash is cash that allows companies to do the same amount of business with fewer employees (cost savings) or handle more business without adding people (cost avoidance).

Soft cash is things such as increased customer satisfaction, reduced time to market, cost avoidance, lost profit avoidance, improved employee morale, enhanced image for the organization, and other intangibles that may result in additional savings to the organization but are harder to quantify.

Each organization may define hard and soft cash differently.

Return on Assets

Return on assets (ROA) is sometimes used for Six Sigma projects to determine whether the use of organization assets is warranted based on the return realized.

The formula for ROA is

$$ROA = \text{Net income/Total assets}$$

When applied to a project, net income refers to expected earnings that result directly from the project's results. Total assets refer to the value of the assets applied to a project.

Return on Investment

Return on investment (ROI) is also used for Six Sigma projects to determine whether the investment in the project is warranted based on the return realized.

The formula for ROI is

$$ROI = \text{Net income/Investment}$$

When applied to a project, net income refers to expected earnings that result directly from the project's results. Investment refers to the value of the outlay made in the project.

Net Present Value

Net present value (NPV) is also known as discounted cash flow (DCF). Calculating the NPV is a method for evaluating projects or investments, and is utilized in capital budgeting. It is an application of a fundamental concept in economics and finance called the time value of money, which, in turn, utilizes the arithmetic of compound interest in reverse. Money received in the future is less valuable than money received today.

An NPV allows the calculation of the current benefit of a project for each window of time and over the total duration of the project for all time periods. If the project NPV is positive, then the project is usually approved for further consideration.

The NPV equation is

$$\text{NPV} = \sum_{i=0}^{n} \frac{CF_t}{(1+r)^t} \tag{3.1}$$

where n is the number of time periods, t is the time period, r is the discount rate or the rate of return, and CF_t is the cash flow in time period t.

Note the following regarding this equation:

- CF_0 is the cash flow in time period zero, which is the same as the initial investment.
- Cash flow for a given time period is calculated by taking the cash flow for project benefits $(CF_{b,t})$ in time period t and subtracting the cash flow for project costs $(CF_{c,t})$ in the same time period.
- i (the annual interest rate) may be substituted for r (rate of return).

Net Present Value: Converting Annual Percentage Rate

To convert an annual percentage rate of i to a rate r for a shorter time period with m time periods per year, use the following equation:

$$r = (1 + i)1/m - 1 \tag{3.2}$$

As an example, a project is planned to update manufacturing equipment and refine the manufacturing process in an automobile parts plant. The cost of capital is 9.5% annual percentage rate (APR).

Project benefits: Cash inflow
Decrease rework and scrap of $700 in month 3
Decrease rework and scrap of $500 in month 4
Decrease rework and scrap of $450 in month 5
Decrease rework and scrap of $450 in month 6

Project costs: Cash outflow
Initial process redesign and training costs $400 in month 1
Installing new equipment costs $840 in month 2
A second round of training costs $100 in month 6

See Table 3.4 for a cash flow summary for this example.
The APR of 9.5% is converted to a monthly rate using Equation 3.2:

$$r = (1 + 0.095)1/12 - 1 = 0.00759$$

The calculation for NPV is

$$NPV = \sum_{i=0}^{n} \frac{CF_t}{(1+r)^t} = \frac{0}{(1+0.00759)^0} + \frac{-400}{(1+0.00759)^1} + \frac{-840}{(1+0.00759)^2}$$

$$+ \frac{700}{(1+0.00759)^3} + \frac{500}{(1+0.00759)^4} + \frac{450}{(1+0.00759)^5} + \frac{450-100}{(1+0.00759)^6}$$

$$= \$712.80$$

Although financial benefits weigh most heavily in the decision to proceed with a Six Sigma project, other benefits also accrue to the organization and may be considered in the decision-making process.

Other benefits to the organization from a successful process improvement project may include

- Improved market position relative to competitors
- Improved ability to meet customer needs, especially enhanced service
- Organization behaviors aligned with vision and values
- Newly created market opportunities
- Project-infused spirit of continuous improvement
- Improved employee morale

TABLE 3.4

Cash Flow Analysis Table

Month	0	1	2	3	4	5	6
Positive cash flow	0	0	0	700	500	450	450
Negative cash flow	0	400	840	0	0	0	100

- Improved overall productivity
- Decreased cycle time
- Increased simplicity (such as for improved customer experiences or ease of manufacture)
- Compliance with existing or emerging regulations or laws

TEAM MANAGEMENT

Obviously teams can outperform individuals. Teams offer more than just increased efficiency. Because the team members are empowered to deal with many things that affect their work, teams provide a great source of job satisfaction and employee involvement.

4

Define

KEY CONCEPTS AND TOOLS

1. Voice of the customer (VOC)
   - Internal and external customers
     - Divide customers for each project and show how the project will impact both internal and external customers.
   - Customer satisfaction
   - Customer value and customer value analysis—survey
   - Quality function deployment (QFD) and critical-to-quality (CTQ) flow-down
     - Define, select, and use appropriate tools to determine customer requirements, such as QFD and CTQ flow-down.
2. Project charter
   - Problem statement
   - Project scope
   - Project goals
   - Project performance measures
3. Project tracking
   - Gantt charts
   - Critical path analysis
   - Program evaluation and review technique (PERT) charts
   - Work breakdown structure

VOICE OF THE CUSTOMER

Internal and External Customers

The concept of internal customers has been widely used as part of the quality movement as a way to break barriers and promote cross-functional cooperation. Slogans like "the customer is the king" are heard in many places of work.

An internal customer is an employee or department who receives an output from another employee or department. Internal customers have also been defined as the next person in the work process.

Based on this concept, all work-related activities of an organization can be viewed as a series of transactions between internal suppliers and internal customers.

External customers are people or organizations who receive and pay for a product. Types of external customers include

End user or final customer: End users or final customers purchase or receive products for their own use and are not an employee or legal part of the supplier organization.

Intermediate customers: Intermediate customers include dealers, distributors, and brokers who buy products and make them available to the final user. Organizations that repair or modify a product for the end user are also generally considered intermediate customers.

Customer–Supplier Chain

Companies understand that business is driven by customer requirements. Many choose a model of two-way interaction between their suppliers, their customers, and themselves (Figure 4.1). This model promotes the exchange of information about requirements and performance.

Customer Satisfaction

A general view of customer satisfaction is meeting customer requirements. Six Sigma emphasizes exceeding customer requirements and thereby "delighting the customer." Customer delight is defined as exceeding

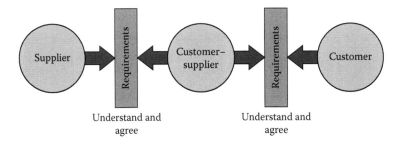

FIGURE 4.1
Customer–supplier chain to understand and agree on mutual requirements.

customer requirements—in ways the customer finds valuable and in ways the customer did not anticipate, but has come to expect afterward.

Customer need can be translated into customer value. Determining customer value is generally a combination of different factors:

- Product attributes
- Product quality
- Service
- Company image and personnel
- Selling price
- Total cost of using product

Customer Satisfaction Survey

This is also called customer value analysis. Customer value analysis is the process many companies use to determine what customers value in terms of the company's and competitors' products.

Customers are asked to

- Identify the product attributes of the product that they value
- Rank the relative importance of these attributes from most to least important
- Assess the company's performance on these attributes
- Rate the attributes of the company's product against those of competitors

Once these questions have been answered, the company can then focus its efforts on improving those product attributes that customers value most.

Quality Function Deployment

QFD is simply the voice of the customer translated into the voice of the engineer. It was developed by Yoji Akao in Japan in 1966. It is also known as "house of quality" (Figure 4.2).

In Akao's words, QFD "is a method for developing a design quality aimed at satisfying the consumer and then translating the consumer's demand into design targets and major quality assurance points to be used throughout the production phase.... (QFD) is a way to assure the design quality while the product is still in the design stage." It has been found that when appropriately applied, QFD results in the reduction of development time by one-half to one-third (Akao 1990).

The three main goals in implementing QFD are

1. Listen to the voice of the customer to prioritize specified and anticipated customer wants and needs. This is a product concept and planning stage. The goal here is to document customer requirements, warranty data, competitive opportunities, product measurements, competing product measures, and the technical ability of the organization to meet each customer requirement. Getting good data from the customer at this stage is critical to the success of the QFD process.

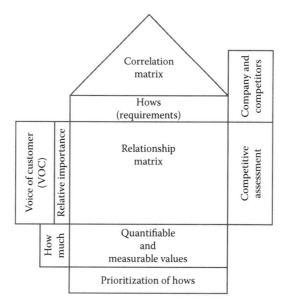

FIGURE 4.2
House of quality.

2. In the correlation matrix, translate these needs into technical characteristics and specifications. This is achieved through process planning led by manufacturing and engineering. During process planning, manufacturing processes are flowcharted and process parameters (or target values) are documented.
3. Benchmark target values to build and deliver a quality product or service by focusing everybody toward customer satisfaction. At this stage, decisions are made about the risks some processes may pose and controls are put in place to prevent risks and failures by the quality assurance department in concert with manufacturing.

Expanded House of Quality

The first phase in the implementation of the quality function deployment process involves putting together a house of quality, such as the one shown in Figure 4.3.

Steps to the house of quality are

Step 1: After negotiations with the customer regarding their requirements, customer requirements are rated according to their preference on a scale from 1 to 5 and tabulated as shown in Figure 4.4 using an Excel spreadsheet.

Step 2: Customer rating of the competition. Understanding how customers rate the competition can be a tremendous competitive advantage. In this step of the QFD process, it is also a good idea to ask customers how your product or service rates in relation to the competition (Figure 4.5).

Step 3: Determine the company's ability to meet customer needs. The relationship matrix is where the team determines the relationship between customer needs and the company's ability to meet those needs (Figure 4.6). The team asks the question, "What is the strength of the relationship between the technical descriptors and the customer's needs?" Relationships can be weak, moderate, or strong and carry a numeric value of 1, 4, or 9.

Step 4: Organizational difficulty. Rate the design attributes in terms of organizational difficulty (Figure 4.7). It is very possible that some attributes are in direct conflict. For example, increasing the number of sizes may be in conflict with the company's stock holding policies. This difficulty is rated on a scale of 1–5, from easy to difficult.

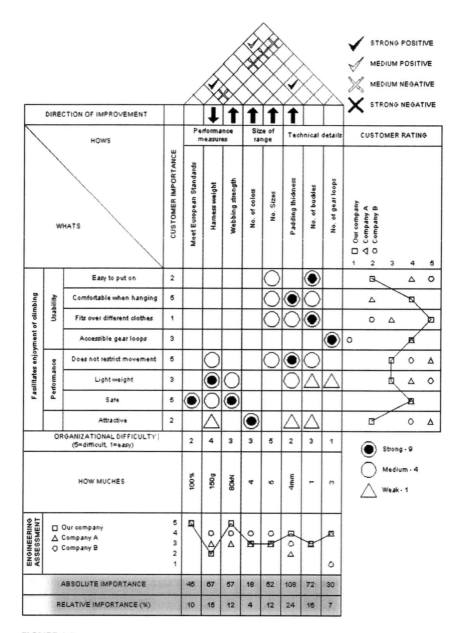

FIGURE 4.3

Expanded house of quality. (From Lowe, A.J., and Ridgway, K., *Quality Function Deployment*, University of Sheffield, UK, 2001, http://www.shef.ac.uk/~ibberson/qfd.html. With permission.)

Facility enjoyment of climbing	Usability	Easy to put on	2										
		Comfortable when hanging	5										
		Fits over different clothes	1										
		Assemble gear loops	3										
	Performance	Does not restrict movement	5										
		Light weight	3										
		Safe	5										
		Attractive	2										

FIGURE 4.4
Preferred customer requirements.

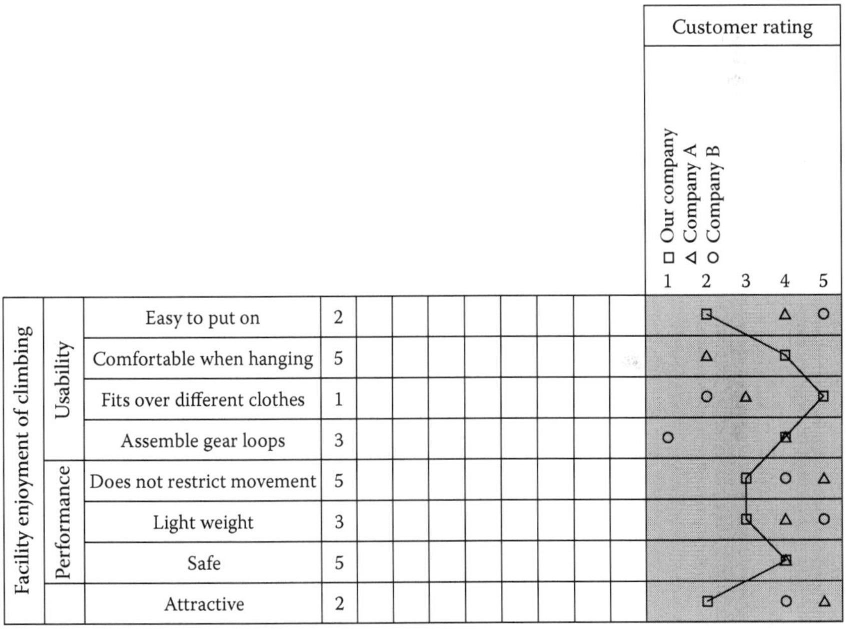

												Customer rating □ Our company △ Company A ○ Company B 1 2 3 4 5
Facility enjoyment of climbing	Usability	Easy to put on	2									□ △ ○
		Comfortable when hanging	5									△ □
		Fits over different clothes	1									○ △ □
		Assemble gear loops	3									○ □
	Performance	Does not restrict movement	5									□ ○ △
		Light weight	3									□ △ ○
		Safe	5									□
		Attractive	2									□ ○ △

FIGURE 4.5
Competitor rating by the customer.

Step 5: Technical analysis of competitor products. To better understand the competition, engineering conducts a comparison of competitor technical descriptors (Figure 4.8). This process involves reverse engineering of competitor products to determine specific values for competitor technical descriptors.

Step 6: Target values for technical specifications (Figure 4.9). At this stage in the process, the QFD team begins to establish target values

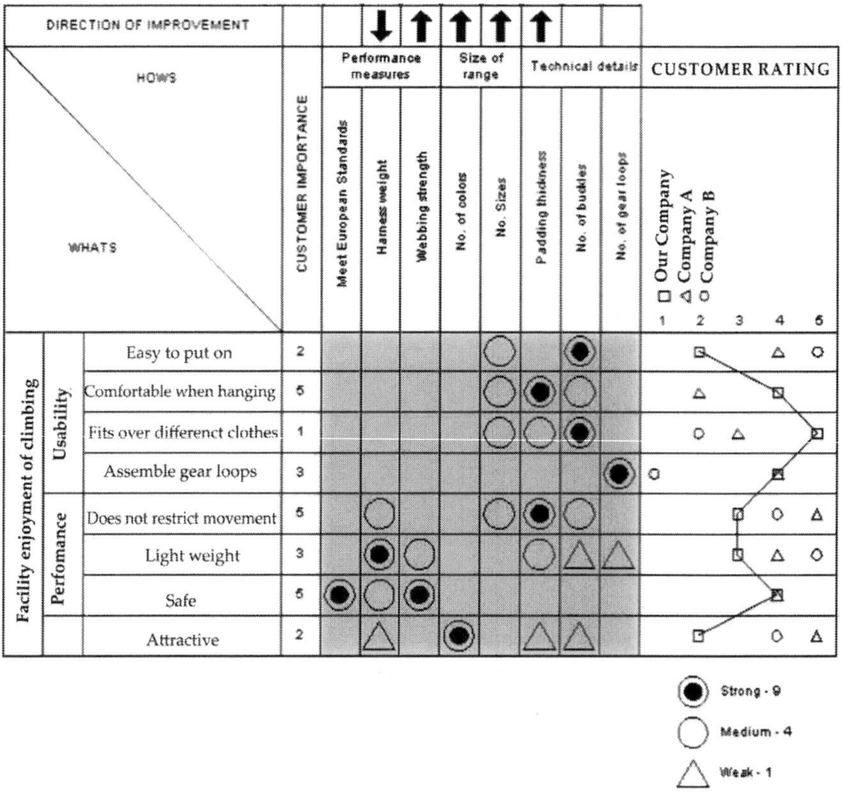

FIGURE 4.6
Determine company's ability to meet customer needs.

and describes the value for each technical specification. Target values represent "how much" for each technical descriptor, and can then act as a baseline to compare against customer requirements and competition.

Step 7: Correlation matrix (Figure 4.10). This stage in the matrix is where the term *house of quality* comes from because it makes the matrix look like a house with a roof. The correlation matrix is probably the least used stage in the house of quality; however, this stage is a big help to the design engineers in the next phase of a comprehensive QFD project. Team members must examine how each of the technical descriptors impact each other. The team should document strong negative relationships between technical descriptors and work to eliminate physical contradictions.

DIRECTION OF IMPROVEMENT	↓	↑	↑	↑	↑			

HOWS

WHATS	CUSTOMER IMPORTANCE	Performance measures			Size of range		Technical details			CUSTOMER RATING
		Meet European Standards	Harness weight	Webbing strength	No. of colors	No. Sizes	Padding thickness	No. of buckles	No. of gear loops	

Facilitates enjoyment of climbing

Usability

WHATS	Imp.	Meet European Standards	Harness weight	Webbing strength	No. of colors	No. Sizes	Padding thickness	No. of buckles	No. of gear loops
Easy to put on	2				○			●	
Comfortable when hanging	5				○	●	○		
Fits over different clothes	1				○	○		●	
Accessible gear loops	3								●

Performance

WHATS	Imp.	Meet European Standards	Harness weight	Webbing strength	No. of colors	No. Sizes	Padding thickness	No. of buckles	No. of gear loops
Does not restrict movement	5		○			○	●	○	
Light weight	3	●	○			○	△	△	
Safe	6	●	○	●					
Attractive	2	△			●		△	△	
ORGANIZATIONAL DIFFICULTY (5=difficult, 1=easy)		2	4	3	3	6	2	3	1

HOW MUCHES

Legend: ● Strong - 9 ○ Medium - 4 △ Weak - 1

Customer Rating: □ Our company ◁ Company A ○ Company B (scale 1–5)

FIGURE 4.7
Rate organizational difficulty to meet customer requirements.

Step 8: Absolute importance (Figure 4.11). Finally, the team calculates the absolute importance for each technical descriptor. This numerical calculation is the product of the cell value and the customer importance rating. Numbers are then added up in their respective columns to determine the importance for each technical descriptor. Now you know which technical aspects of your product matter the most to your customer. You have now reasonably satisfied that you have heard your customer's voice.

We have eight "hows" (company ability to meet customer needs), as shown in Table 4.1.

We have eight "whats" (customer requirements, each with its own customer importance), as shown in Table 4.2.

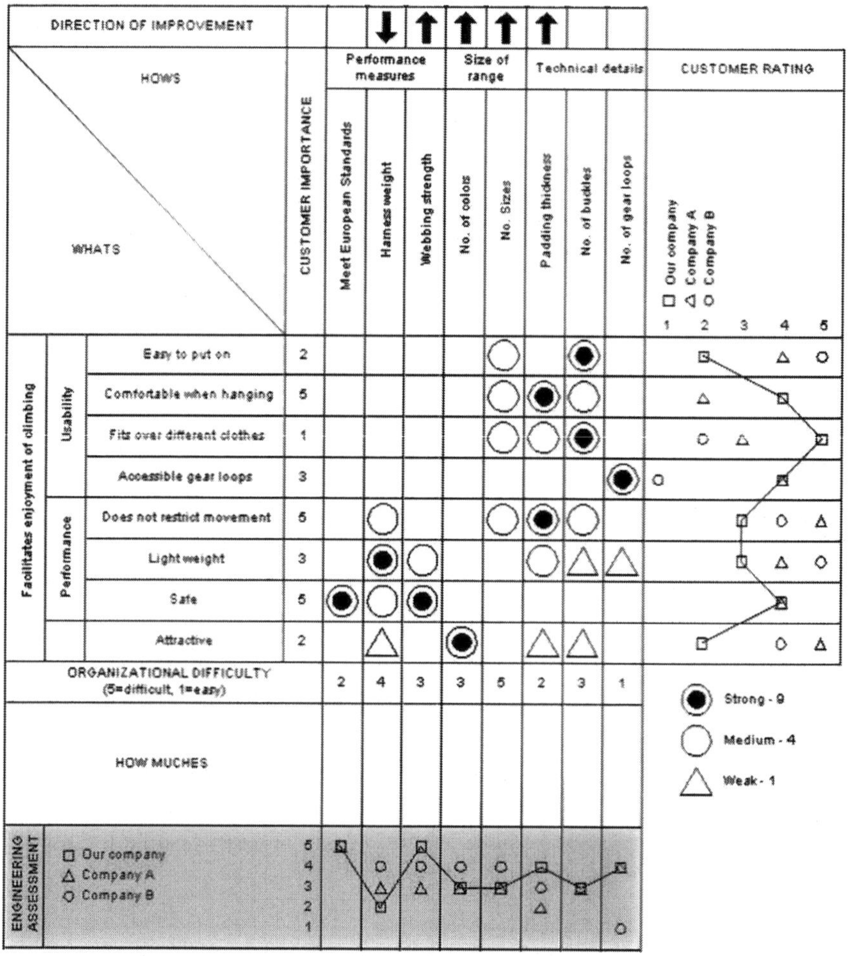

FIGURE 4.8
Technical analysis of competitor products.

Now refer to Figure 4.6.

This relationship matrix is where the team determines the relationship between customer needs and the company's ability to meet those needs. The team asks the question, "What is the strength of the relationship between the technical descriptors and the customer's needs?" Relationships can be weak, moderate, or strong and carry a numeric value of 1, 4, or 9.

In Figure 4.8, the cell value for the customer requirement column "no. of colors" is given as strong, having a value of 9, and the customer importance is 2. So the absolute importance comes to $2 \times 9 = 18$.

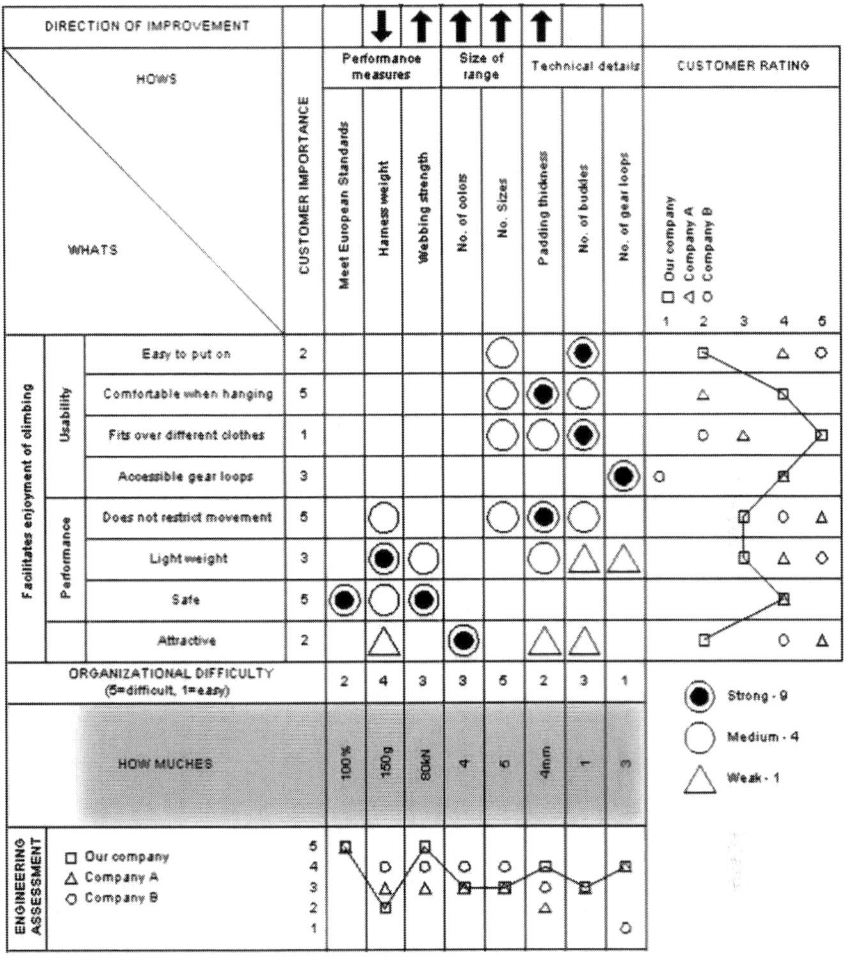

| DIRECTION OF IMPROVEMENT | | ↓ | ↑ | ↑ | ↑ | ↑ | | | | |

HOWS — Performance measures | Size of range | Technical details | CUSTOMER RATING

WHATS / Facilitates enjoyment of climbing

	CUSTOMER IMPORTANCE	Meet European Standards	Harness weight	Webbing strength	No. of colors	No. Sizes	Padding thickness	No. of buckles	No. of gear loops	CUSTOMER RATING
Usability										
Easy to put on	2				○			◉		□ ... △ ○ (2–4)
Comfortable when hanging	5				○	◉	○			△ ... □
Fits over different clothes	1				○	○	◉			○ △ ... □
Accessible gear loops	3							◉	○	□
Performance										
Does not restrict movement	5	○			○	◉	○			□ ○ △
Light weight	3	◉	○		○	△	△			□ △ ○
Safe	5	◉	○	◉						□
Attractive	2	△		◉	△	△				□ ○ △

| ORGANIZATIONAL DIFFICULTY (5=difficult, 1=easy) | 2 | 4 | 3 | 3 | 5 | 2 | 3 | 1 |
| HOW MUCHES | 100% | 150g | 80kN | 4 | 5 | 4mm | 1 | 3 |

Key: ◉ Strong - 9 ○ Medium - 4 △ Weak - 1

ENGINEERING ASSESSMENT — □ Our company, △ Company A, ○ Company B (scale 1–5)

FIGURE 4.9
Target values for technical specifications.

In the same way for the customer requirement "no. of sizes," the cell value given is 4. So for this requirement, the absolute requirement is $(2 \times 4) + (5 \times 4) + (1 \times 4) + (5 \times 4) = 52$.

The relative importance of the company's ability to meet the related customer requirement is arrived at by adding all absolute importance figures $(18 + 52 + \ldots = 452)$ and calling the sum (452) 100%. The relative percentage for "no. of colors" comes to $(1/452\% = 4\%)$, and so on, as shown in Table 4.1.

The QFD exercise is complete.

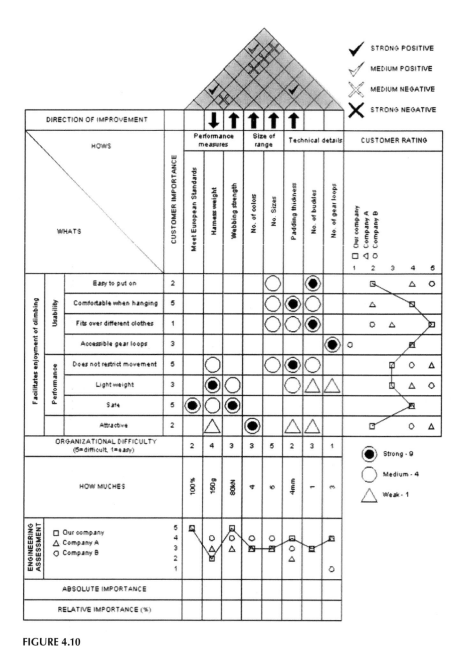

FIGURE 4.10
Correlation matrix.

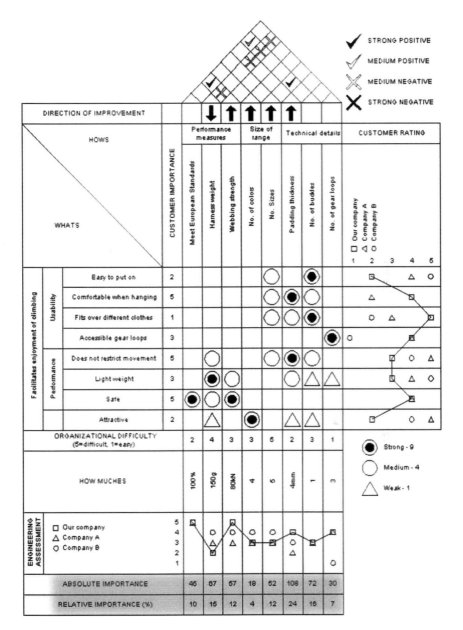

FIGURE 4.11
Absolute and relative importance.

TABLE 4.1

Eight Hows (Company Ability to Meet Customer Needs)

Ability No.	Company Ability to Meet Customer Needs	Absolute Importance	Relative Importance %
1	Meet European standards	45	16
2	Harness weight	67	10
3	Webbing strength	57	14
4	No. of colors	18	4
5	No. of sizes	52	12
6	Padding thickness	108	24
7	No. of buckles	72	16
8	No. of gear loops	30	8

TABLE 4.2

Eight Whats (Customer Requirements)

Requirement No.	Requirement	Importance
1	Easy to put on	2
2	Comfortable when hanging	5
3	Fits over different clothes	1
4	Assemble gear loops	3
5	Does not restrict movement	5
6	Lightweight	3
7	Safe	5
8	Attractive	2

Next Stage

The above process is repeated in a slightly simplified way for the next three project phases.

A simplified matrix involving steps 1–3 and 5–8 above is developed where

- The voice of the engineer is translated into the voice of the part design specifications.
- The design specifications get translated into the voice of manufacturing planning.
- The voice of manufacturing is translated into the voice of production planning.

QFD is a systematic means of ensuring that customer requirements are accurately translated into the technical goals and specifications throughout each stage of product development. The QFD process does not just maintain or improve product performance; it enables organizations to meet or exceed customer demands and delights customers by fulfilling their unarticulated and unanticipated desires and wants.

CTQ Flow-Down

Customer satisfaction generally falls into four dimensions:

- Quality
- Delivery
- Cost
- Safety (for internal customers as well)

Critical to X (CTX) Requirements

This is how we define and describe various CTX requirements. Determine

- Critical to quality (CTQ)
- Critical to cost (CTC)
- Critical to process (CTP)
- Critical to safety (CTS)
- Critical to delivery (CTD)

These requirements should be measurable because they provide important goals and milestones for aligning projects with those requirements.

Cost and delivery are easy to quantify. Customers are willing to pay X money per item and expect its delivery Y days after the order is placed. Quantifying quality characteristics presents more of a challenge, and one of the tools devised to help is called the CTQ flow-down. Its purpose is to start with the high-level organizational strategic goal of customer satisfaction and determine how this goal "flows down" into measurable goals. The nomenclatures for the various levels are illustrated in Figure 4.12.

CTQ Flow-Down Example

The collected data for the result tracker, key support measure, key performance indicator, and key business driver are evaluated using a Pareto

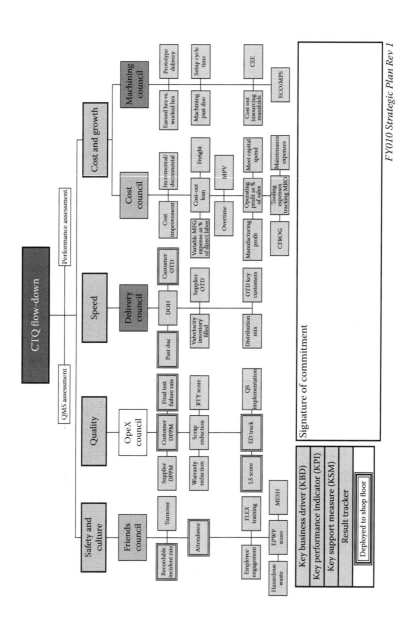

FIGURE 4.12
CTQ flow-down.

chart to obtain at least one major CTQ measure that can be measured and a process sigma that can be calculated.

Use the following tools to collect data at the Define stage (the same tools can be used at the Analyze stage):

Box and whisker plot: A tool used to display and analyze multiple sets of variation data on a single graph.

Check sheet: A generic tool that can be adapted for a wide variety of purposes. The check sheet is a structured, prepared form for collecting and analyzing data.

Control chart: A graph used to study how a process changes over time. Comparing current data to historical control limits leads to conclusions about whether the process variation is consistent (in control) or unpredictable (out of control, affected by special causes of variations).

Design of experiments: A method for carrying out carefully planned experiments on a process. Usually, a design of experiments involves a series of experiments that start by looking broadly at a great many variables and then focuses on the few critical ones.

Histogram: The most commonly used graph for showing frequency distributions, or how often each different value in a set of data occurs.

Scatter diagram: A diagram that graphs pairs of numerical data, one variable on each axis, to look for a relationship.

Stratification: A technique that separates data gathered from a variety of sources so that patterns can be seen.

Survey: Data collected from targeted groups of people about their opinions, behavior, or knowledge.

PROJECT CHARTER

Project Statement

A project statement starts with a business case first, which leads to the problem statement.

Business Case

The business case is a short summary of the strategic reasons for the project (Table 4.3). The justification for a business case would typically involve at least one of the following:

- Cost per unit
- Quality or defect rate
- Cycle time

The problem statement will detail the issue that the project team wants to improve (Table 4.4). Problem cases include

- Historical data
- What areas of the business are affected
- How long the problem has existed
- Any other symptoms of the problem

Project Scope

The project scope is the specific aspect of the problem that will be addressed, and it serves to specify the boundaries of the project.

Project Goal

Goals should

- Be carefully thought out and expressed.
- Specify how completing the project will lead to improvements over the existing state of affairs. You should be able to clearly describe the outcomes, deliverables, and benefits to stakeholders and customers.
- Provide the criteria you need to evaluate the success of the project in terms of time, costs, and resources.
- Be reviewed by the core team, which must reach consensus before moving to the next phase of the project.

TABLE 4.3

Example of Business Case Summary

Current Situation	Improvement Objective	Key Business Driver	Performance Category	Metric	Current 2013	Goal Q3-13	Goal Q3-14
Current logistic process causes 50% of finished goods to reach our OEMs and direct customers late	Improve OTD	Customer satisfaction	Customer satisfaction	OTD	72%	80%	90%
Missing and lost shipments	OTD	Customer satisfaction	Customer satisfaction	Zero missed shipments	72%	80%	90%
The cost-out opportunities due to line down costs, credit to customers, and premium freight	Manufacturing profit	Business excellence	Cost improvement	Cost-out	$400K lost so far in 2013	$25K	$75K
Incoming raw material and customer returned products arriving late to the plants and also getting lost in Pharr Warehouse	Manufacturing profit/OTD	Business excellence Customer satisfaction	Cost improvement	Cost-out	$150K lost so far in 2013	$25K	$50K
Pharr Warehouse does not have adequate resources (system and management) to report Swiss company transactions	Manufacturing profit/OTD	Business excellence Customer satisfaction	Cost improvement	Cost-out	$40K lost so far in 2013	$25K	$75K

Note: OEM, original equipment manufacturer; OTD, on-time delivery.

TABLE 4.4

Poor and Well-Written Problem Statements

Problem Statement Example	
Poorly Written Problem Statements	**Well-Written Problem Statements**
There are too many customer returns. The hardware department return rate is 17%.	In 2012, the return rate was 17%, representing $15 million in returns. This was 7% higher than the target objective and goal for the division.
There are too many incorrect customer invoices. We must reduce incorrect invoices by 15%.	In the fourth quarter, 20% of all customer invoices were incorrect. This was an increase of 5% from the third quarter.

SMART Goals

The acronym frequently used to assess whether a project's goals are good is SMART: specific, measurable, attainable, relevant, and time-bound. These concepts are illustrated below:

Specific: Clear, appropriately narrow, and easy to describe

Measure: Has a specific outcome and can be measured easily

Attainable: Suggests the need for significant improvement in the process, but is not impossible

Relevant: Has strategic significance to the customer and company

Time-bound: Has an end date where metrics can be measured to determine the success

Project Performance

Project performance reviews are meetings between the project leader (typically the Black Belt) and the management team. They allow the project leader to update management on the status of the project in terms of financial progress.

In some organizations, decision gates (toll gates) are both an opportunity to review the project with senior management and an opportunity to update key stakeholders on the project's overall progress (i.e., its current phase) (Figures 4.13 through 4.15).

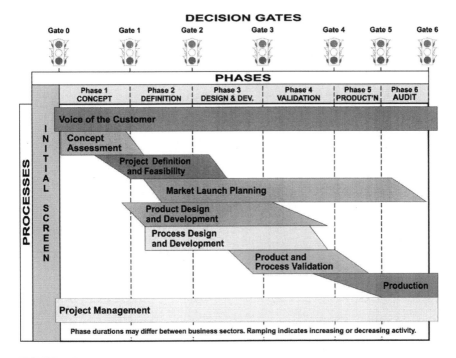

FIGURE 4.13
Project phases, decision gates, and performance review map.

Project Performance Metrics

Performance metrics fall into three categories:

- Quality
- Time
- Cost

These three metrics work within a delicate balance. Improving quality might require an increase in the time needed for production, which would also increase costs. Reducing the cycle time might reduce costs, but it also might reduce quality. Reducing costs might also reduce quality if source materials are less expensive because they are inferior in quality.

Quality metrics include

- Fewer defects
- Higher quality of material resources

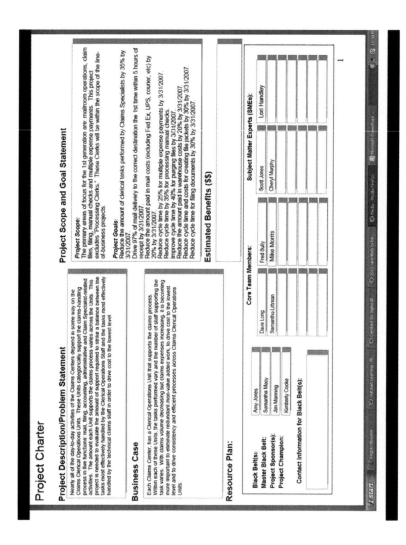

FIGURE 4.14
Project charter.

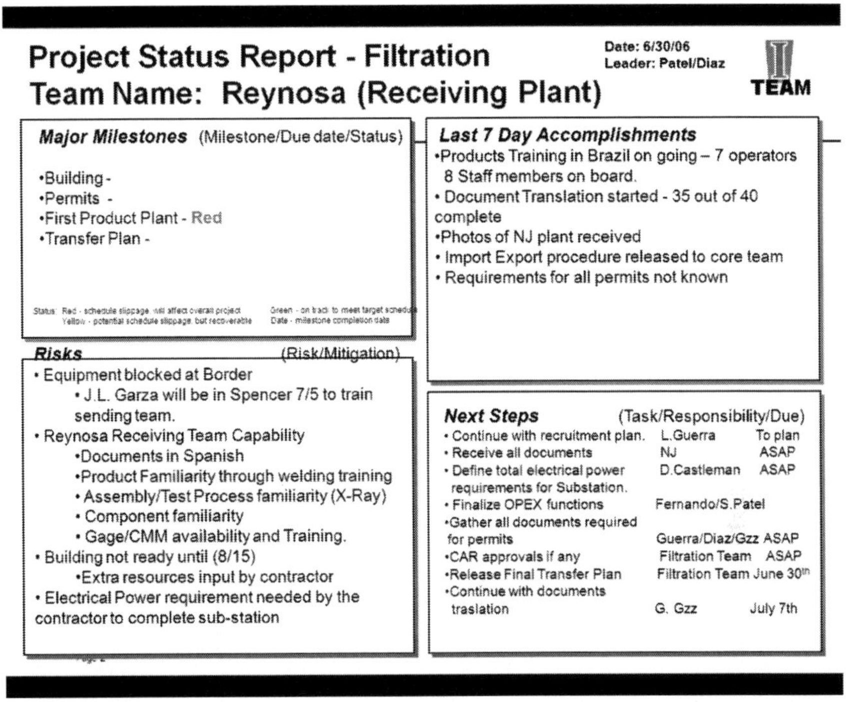

Project Status Report - Filtration
Team Name: Reynosa (Receiving Plant)

Date: 6/30/06
Leader: Patel/Diaz

TEAM

Major Milestones (Milestone/Due date/Status)

•Building -
•Permits -
•First Product Plant - Red
•Transfer Plan -

Status: Red - schedule slippage -will affect overall project   Green - on track to meet target schedule
Yellow - potential schedule slippage, but recoverable   Date - milestone completion date

Last 7 Day Accomplishments
•Products Training in Brazil on going – 7 operators
8 Staff members on board.
• Document Translation started - 35 out of 40
complete
•Photos of NJ plant received
• Import Export procedure released to core team
• Requirements for all permits not known

Risks (Risk/Mitigation)
• Equipment blocked at Border
    • J.L. Garza will be in Spencer 7/5 to train
    sending team.
• Reynosa Receiving Team Capability
    •Documents in Spanish
    •Product Familiarity through welding training
    • Assembly/Test Process familiarity (X-Ray)
    • Component familiarity
    • Gage/CMM availability and Training.
• Building not ready until (8/15)
    •Extra resources input by contractor
• Electrical Power requirement needed by the
contractor to complete sub-station

Next Steps (Task/Responsibility/Due)
• Continue with recruitment plan. L.Guerra To plan
• Receive all documents NJ ASAP
• Define total electrical power D.Castleman ASAP
  requirements for Substation.
• Finalize OPEX functions Fernando/S.Patel
•Gather all documents required
for permits Guerra/Diaz/Gzz ASAP
•CAR approvals if any Filtration Team ASAP
•Release Final Transfer Plan Filtration Team June 30th
•Continue with documents
traslation G. Gzz July 7th

FIGURE 4.15
Project status report (4-up chart).

- Fewer warranty items or returns
- Higher demand by customers for product or service
- Customer survey scores increasing

Time metrics include

- Cycle time reductions
- Response time in a call center
- Time to market
- Response time to customer inquiries
- Time to complete special orders

Cost metrics include

- Revenue realized due to increased sales of a product or service that, in turn, is due to a lowered price
- Reduced costs of production and volume improvement

- Improved product quality and enhanced product features
- Better availability to the customer, fewer defects, and so on
- Cost reductions realized through fewer defects, less scrap, fewer returns, fewer warranty items, and so on
- Many companies call this the cost-out data, which are compiled after each improvement project is completed

---

PROJECT TRACKING

Some of the project planning and tracking tools with which one must be familiar are

- Gantt chart
- Critical path analysis
- PERT chart

Gantt Chart

The Gantt chart was developed by Henry Gantt in the 1910s (Figure 4.16). "A Gantt chart is a horizontal bar chart that shows the tasks of a project, when each must take place, and how long each will take. As the project progresses, bars are shaded to show which tasks have been completed." (Tague 2005).

Gantt Chart Procedure

Steps for creating a Gantt chart include

1. Identify the following tasks needed to complete the project:
   a. Key milestones (important check points)
   b. Time required for each task
   c. The sequence
      - Tasks to be finished before the next task can begin
      - Simultaneous tasks
      - Tasks to be completed before each milestone
2. Draw a horizontal time axis along the top or bottom of a page, and then mark off in an appropriate scale for the length of the tasks (days or weeks).

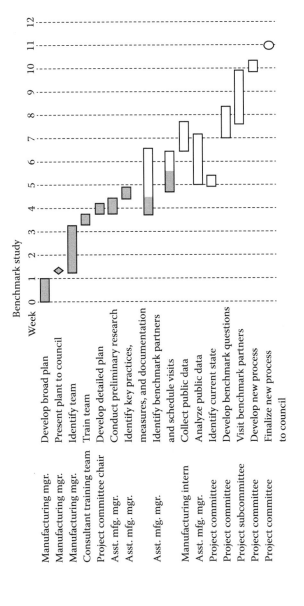

FIGURE 4.16
Gantt chart. (From Teague, N.R., *The Quality Toolbox*, 2nd ed., American Society for Quality, Caguas, Puerto Rico, 2005.)

3. Down the left side of the page, write each project task and milestone in order:
   a. For each event happening at a point in time, draw a diamond (unfilled) under the time for that event.
   b. For activities occurring over a period of time, draw a bar (unfilled) spanning the appropriate time.
      - Align the left end of the bar with the time the activity begins.
      - Align the right end with the time the activity concludes.
      - Ensure that every task of the project is on the chart.
4. As events and activities take place, shade the diamonds and bars to show completion. For tasks in progress, estimate the percentage of completion and shade the appropriate amount.
5. Place a vertical marker to show the current date. When posting the chart on the wall, a heavy dark string hung vertically across the chart with two thumbtacks is an easy way to show the current time.

Critical Path Analysis

Developed by project managers in the 1950s, critical path analysis (CPA), also known as the critical path method (CPM), is one of several planning tools for demonstrating and viewing chronological tasks, identifying possible timing risks, and establishing the least amount of time for the project or process (Figure 4.17).

The benefits of CPA are that it

- Displays a graphical model of the project
- Predicts the shortest time required to complete the project

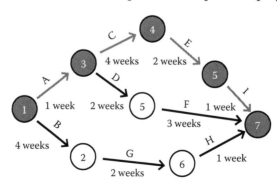

FIGURE 4.17
Critical path analysis.

- Emphasizes activities critical to maintaining the schedule
- Provides a timing reference point throughout the project
- Identifies interrelationships between tasks

Three paths exist in Figure 4.17:

1-3-4-5-7 (8 weeks) (critical path, longest time)
1-3-5-7 (6 weeks)
1-2-6-7 (7 weeks)

Critical Path Analysis Procedure

1. Use Post-it notes to individually list all tasks in the project. Underneath each task, draw a horizontal arrow pointing right.
2. Arrange the notes in the appropriate sequence (from left to right).
3. Between each pair of tasks, draw circles to represent events and mark the beginning or end of a task.
4. Label all events, in sequence, with numbers. Label all tasks, in sequence, with letters.
5. For each task, estimate the completion time. Write the time below the arrow for each task.
6. Draw the critical path by highlighting the longest path from the beginning to the end of the project.

PERT Chart

First developed by the Navy in the 1950s to manage complex projects, program evaluation and review technique charts are powerful tools for reducing the time and cost for completing a project. PERT's planning technique accounts for randomness in time requirements (Figure 4.18).

The term *critical path method* will also be used for a similar approach. "The most important difference between PERT and CPM is that originally the time estimates for the activities were assumed deterministic in CPM. Deterministic methods use point estimates which are often, but not necessarily always, worst case estimates. The probabilistic estimates in PERT methods allow for the use of distributions to represent the long-term profiles, and distributions of each estimate. Today, PERT and CPM actually comprise one technique and the differences are mainly historical" (Pyzdek and Keller 2011).

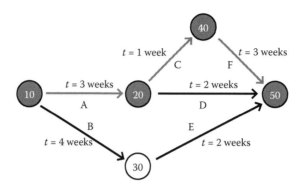

FIGURE 4.18
PERT chart.

Time-dependent task diagrams allow for the plotting of each task in relation to the other. CPM charts add the dimension of normal (most likely) time to complete tasks, and the critical path (longest timeline) through the project.

The benefits of the PERT chart are that it

- Provides a means of estimating the project completion time
- Demonstrates the probability for completing the project ahead of schedule
- Identifies start and end dates, as well as critical path activities that affect completion time
- Organizes tasks in established time frames
- Acts as a decision-making tool
- Identifies where and when parallel activities occur
- Serves as an evaluation tool to determine the effect of changes

PERT Chart Example

There are five processes that can be performed for the compliance or audit process. Create the report "10" and send it to electronic storage "20," which will take three weeks to be received. A determination of criticality is made. If it is critical, then it must be delivered to audit to be validated for accuracy by an individual "40," which takes one week. The validation process will occur and it will be sent to internal audit "50" in three weeks to be placed in queue for quarterly audits. If the determination "20" is not critical, then it is sent in two weeks to be placed in queue for quarterly audits by internal audit "50." If the report is sent to physical storage "30," it will take four weeks to be processed, and a random group is sent to internal audit "50," which will take two weeks.

The quickest path to the audit queue is the noncritical electronic report that will take five weeks to be moved into the internal audit queue. The longest path is the electronic critical reports, which will take seven weeks to be moved into the internal audit queue.

PERT Chart Key Terms

The following terms are important to keep in mind when using PERT charts. *Critical path* refers to the sequence of tasks (path) that takes the longest time and determines the project's completion date. Any delay of tasks on a critical path will delay the project completion time.

Slack time refers to the time an activity can be delayed without delaying the entire project. Tasks on the critical path have zero slack time. Slack time is the difference between the latest allowable date and the earliest expected date. It is represented by the following:

$$\text{Slack time} = T/L - T/E$$

where T/E is the earliest time (date) on which an event can be expected to occur and T/L is the latest time (date) on which an event can occur without extending the completion date of the project.

An *event* is the starting or ending point for a group of tasks.

Activity is the work required to proceed from one event or point in time to another.

PERT Chart Procedure

1. Use Post-it notes to individually list all tasks in the project. Underneath each task, draw a horizontal arrow pointing right.
2. Arrange the notes in the appropriate sequence (from left to right).
3. Between each two tasks, draw circles to represent events and to mark the beginning or end of a task.
4. Label all events, in sequence, with numbers. Label all tasks, in sequence, with letters.

5. For each activity, make three estimates regarding time requirements: the shortest possible time, the most likely time, and the longest time.
6. Determine the critical path.
7. Adjustments to the PERT chart should be made to reflect any changes to the project along the way.

DEFINE PHASE SUMMARY

The purpose of the define phase is to

- Establish the specific Six Sigma project scope and boundaries
- Define the specific business problem
- Describe the business process needing improvement
- State the goals to be achieved

5

Measure

The second phase of the define, measure, analyze, improve, and control (DMAIC) process is measure. It is a measure of how the key business processes are doing. It is said that if it is not measured, it is not managed. A business process consists of a series of events to produce an output. The objective is to add value at each event to produce a product or service that in turn can be sold to the customer.

Major concepts and tools discussed in the measure phase are

- Process characteristics
- Measurement system analysis (MSA)
- Process capability measurement

PROCESS CHARACTERISTICS

Process characteristics have an origin with written documentation (Figure 5.1). First it is necessary to review the written procedures, work instructions, flowcharts, and quality records in order to determine process characteristics.

A process map is then created for each process where all steps of the process actions are enumerated. These process maps are used for all measure activities.

Process characteristics comprise the following input and output variables:

- Suppliers, inputs, process, outputs, and customers (SIPOC)
- Work in progress (WIP)
- Work in queue (WIQ)
- Touch time

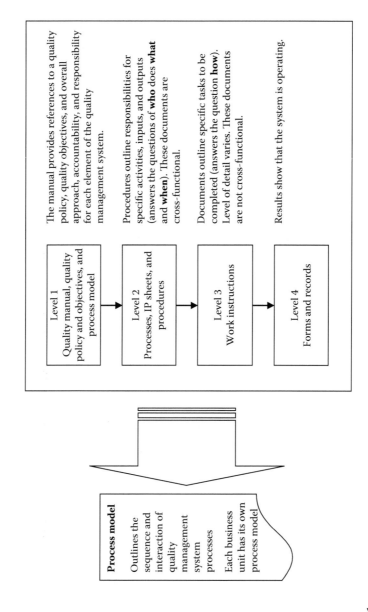

FIGURE 5.1

Quality system documents and their hierarchy.

- Takt time
- Cycle time
- Throughput
- Value stream maps
- Process maps
- Flowcharts
- Procedures
- Work instructions
- Spaghetti diagrams
- Circle diagrams

We have defined the processes through input, process, and output variables, as well as using the turtle and SIPOC diagrams in Book 1. The relationships of these variables and their defined characteristics need to be measured for process improvement.

Identify processes having wastes and poor utilization. Improve these processes by analyzing work in progress, work in queue, touch time, takt time, cycle time, throughput, and so forth.

Analyze processes by preparing value stream maps, process maps, flowcharts, procedures, work instructions, spaghetti diagrams, circle diagrams, and so forth.

These variables are discussed below.

SIPOC: Covered in Books 1 and 2.

WIP: Incomplete product or service that is awaiting further processing and that has not reached the output stage. Surprisingly, accounting systems consider WIP an asset, but it is in many ways a liability because it needs upkeep expenses such as storage, environmental control, identifications, and record keeping.

WIQ: Material waiting in queue and is a component of WIP. It is important to identify WIQ at each step of the process to identify process bottlenecks or constraints. The first-in, first-out (FIFO) policy helps in reducing WIQ as long as necessary attention is given to the shelf life of the inventory.

Touch time: Time that material is actually being touched or processed at one of the steps. The processes with large touch times can be identified for improvement purposes.

Takt time: Covered in Book 2.

Cycle time: Covered in Book 2.

Throughput: Covered in Book 2.

Value stream map: Covered in Book 2.

Process map: Covered in Book 2.

Flowchart: Covered in Book 2.

Procedures: Covered in Book 2.

Work instructions: Covered in Book 2.

Spaghetti diagram: Drawing that is a snapshot in time showing the layout and flow of materials (product flow), information (paper flow), and people (people flow) in a work area (Figure 5.2).

A spaghetti diagram uncovers motion and transportation wastes.

Circle diagram: Type of graphic organizer that helps you find general information about a character that you can organize and interpret later. For example, see the Six Sigma benefits diagram in Figure 5.3.

One other important concept in the business process concerning input/output variables is that of process handoffs.

Process Handoff Diagram

Many business processes have to pass through multiple departments. An example of such a process is quote-to-cash, shown in Figure 5.4. The diagram shows the process from one department to the next, demonstrating how each transition is a potential source for waste or inefficiency. At any of these points, data get lost, waste is created, delays mount up, information is lost, and the process becomes less effective and less efficient.

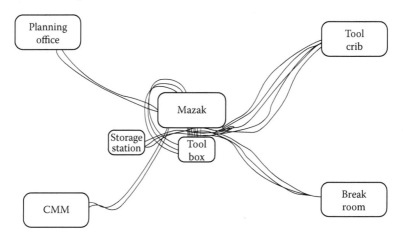

FIGURE 5.2
Spaghetti diagram for a machine shop.

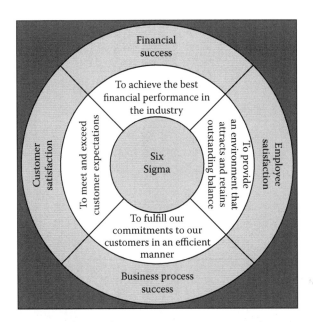

FIGURE 5.3
Circle diagram for Six Sigma benefits.

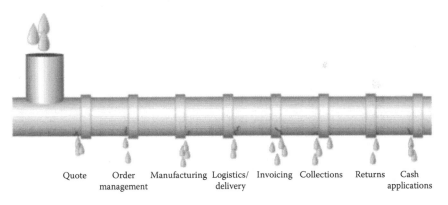

FIGURE 5.4
Process handoff diagram.

The opportunities to review, identify defects, and improve the process lie at each one of the following handoffs:

- Quote
- Order management
- Manufacturing
- Logistics/delivery

- Invoicing
- Collections
- Cash applications

Challenges at Cross-Functional Areas

It can be challenging and difficult from a human management perspective to work toward improving efficiencies across departments. Here is what may happen at a typical management review meeting:

- Department leaders normally overlook the common handoff problem and bring to these meetings their own agendas and behavior based on incentives.
- It may happen that this is the first time two department leaders are in a meeting together working toward a common goal.
- Furthermore, the nature of the existing processes may have created a relationship that is antagonistic between the two departments. Making such processes efficient takes a great deal of cooperation and often major change.
- Barriers need to be broken (Figure 5.5).

Data Collection

Measure what is measurable, and make measurable what is not so.

Galileo Galilee

In God only we trust, everyone else please bring data.

Anonymous

FIGURE 5.5
"Breaking the Barriers" cartoon. (Courtesy Pareto Head, ASQ Quality Progress, April 2013.)

A quality system produces process data that are measured and recorded in quality records.

Measured data not only guide the process improvement, but also forewarn, inform, and guard an organization against warranty and liability costs.

Qualitative and Quantitative Data

Qualitative data
- Deals with descriptions
- Can be observed but not measured, for example, colors, textures, smells, tastes, appearance, and beauty

Quantitative data
- Deals with numbers
- Can be measured, for example, length, height, area, volume, weight, speed, time, temperature, humidity, sound level, cost, membership, and age

Continuous (Variables) and Discrete (Attributes) Data

Discrete data
- Counted, for example, 30 students in a class (one cannot have half a student) or 25 apples in a case

Measurement Scales

Measurement scales are used to quantify variables. Table 5.1 describes the four scales of measurement that are commonly used in statistical analysis:

1. Nominal scales
2. Ordinal scales
3. Interval scales
4. Ratio scales

Continuous data
- Also called variable data
- Measured on a continuum or scale
- Can be subdivided into smaller measurements limited only by the measurement system
- Possibly has a decimal for more accuracy

TABLE 5.1

Measurement Scales

Scale Type	Definition	Example	Calculations
Nominal	Label indicating a characteristic by a name, or presence/absence. Can only count items.	Pass or fail, accept or reject, male or female, 0 = red, 1 = white, 2 = blue	Proportion, chi-square
Ordinal	Grouping into categories having an attribute. When comparing elements, order is important.	Likert scale and 1–5 scales	Rank order correlations
Interval	Fixed or defined scale differentiating two successive points, but no true 0. Potential zero point in measurements.	Time of day (calendar time), temperature	Correlations T-tests F-tests Regressions
Ratio	Relationship between two values. Characteristic able to be measured; there is a true 0. Zero point indicates an absence of an attribute. Can add, subtract, multiply, and divide these values.	% purity, miles per hour, height in inches, elapsed time	Correlations T-tests F-tests Regressions

- Generally more powerful than discrete data because they can be analyzed statistically
- Examples are
  - Physical measurements (length, volume, width, time, temperature, etc.)
  - Using a rule to measure a single dimension only to the nearest 0.1 inch
  - Using a venire caliper to measure the same dimension to the nearest 0.005 inch
  - Using a micrometer to measure the same dimension to the nearest 0.0001 inch

MEASUREMENT SYSTEM ANALYSIS

Business uses of MSA include the following:

- It is a mandatory requirement for ISO TS 16949 and AS 9100 certification.
- It discovers potential sources of process variation.

- It minimizes defects due to measurement errors.
- It improves product quality and enhances customer satisfaction.

The purpose of MSA is to statistically verify that existing measurement systems (consisting of measurement equipment, the samples to be measured, and the operators who carry out measurements) provide

- Ideal (nearest to the standard) values of the characteristic being measured
- Results without bias
- Least variation

Often, the measurements are not representative of the true value of the characteristic being measured. This might be because the measurement system is

- Not accurate enough—it introduces bias into the measurement
- Not precise enough
- Used improperly by an untrained operator

The measurement system consists of

- Measurement equipment
- Measurement process
  The measurement process basically includes the appraiser (operator) who takes the measurements and the actual process of measuring involving a part or a process.

Sampling Strategy

Sampling strategies are of two types:

1. Probability sampling strategy, where a sample has a known probability of being selected.
2. Nonprobability sampling, where a sample does not have a known probability of being selected. Nonprobability sampling methods are based on human choice and not on random selection, and thus they can have major sources of bias.

In this chapter, we will deal with probability sampling strategy only. Table 5.2 below describes main sampling strategies.

TABLE 5.2

Sampling Strategies

Sampling Strategy	Definition	Examples
Random	Select sample units so that all units have the same probability of being selected. Every unit (n) has an equal chance of being selected for the sample.	Random number generator is used for census taking. Random number generator is used by a fast food chain to print a survey number on a receipt for customers to use when they call and answer questions about their experience.
Systematic	Every nth record is selected from a list of the population. As long as the list does not contain any hidden order, this strategy is just as good as random sampling.	A customer service call center receives a randomly generated list of customers who have interacted with customer service representatives. The manager selects every fifth name on the list to conduct a follow-up call for quality assurance.
Stratified	If the population has identified categories or strata that have a common characteristic, random sampling is used to select a sufficient number of units from each stratum. Stratified sampling is often used to reduce sampling error.	Customer survey results are divided into multiple strata based on gender and income level. A manufacturer divides data on defects into strata based on manufacturing location and equipment type used for production.

Measurement Equipment

The following are types of measurement equipment:

- Attribute screens (go/no-go screens to sort out mined mineral lumps, etc.)
- Gage blocks
- Calipers
- Optical comparator
- Coordinate measuring machine (CMM)
- Micrometer
- Measuring machines for tensile strength, compression strength, and so forth
- Titration (Volumetric conical flask, burette having markings to measure volume and pipette with a volume noted on it.)

Every process has variation. One component of this variation could be the error caused by the measurement system. Measurement system analysis determines that portion of the total process variation that is caused by the measurement system. This way, we are able to see how much the measurement system contributes to overall process variability.

Terms and Definitions

Bias

Bias is the difference between the observed average and the reference (Figure 5.6). If you measure something several times and all values are close, they may all be wrong if there is a bias. Bias is a systematic (built-in) error that makes all measurements wrong by a certain amount.

Examples of bias are:

- The weighing scale reads 0.5 kg when there is nothing on it.
- Your height is measured while you are wearing 1-inch high heels.
- A 1-liter measuring cylinder has a mark at the wrong place.

The equation for bias is

$$\text{Bias} = \frac{\sum_{i=1}^{N} X_i}{n} - T \tag{5.1}$$

where n is the number of times the standard is measured, X_i is the ith measurement, and T is the true value of the standard.

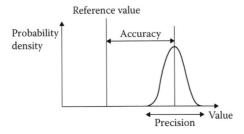

FIGURE 5.6
Bias.

Bias can be removed by calibration.

Two characteristics contribute to the effectiveness of a measurement method:

1. Precision
2. Accuracy

Precision has two components:

- Repeatability: The same person cannot get the same result twice on the same subject.
- Reproducibility: Different people cannot agree on the result obtained on the same subject (part or process).

Gage repeatability and reproducibility (R&R) is the combined estimate of repeatability and reproducibility.

Accuracy is related to the difference between the average and the reference value (this difference is called bias). Accuracy is how close a measured value is to the *actual (true) value*.

The degree of accuracy depends on the instrument you are measuring with (Figure 5.7). But as a general rule, the degree of accuracy is *half a unit* on each side of the unit of measure.

If your instrument measures in 1-inch increments, then any value between 6½ and 7½ is measured as 7 inches.

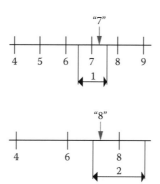

FIGURE 5.7
Degree of accuracy.

If your instrument measures in 2-inch increments, then any value between 7 and 9 is measured as 8 inches.

Precision

As stated above, precision is a total gage variance made up of two components: repeatability and reproducibility (Figure 5.8).

Repeatability (equipment variation) is the inherent variation in measurements taken by a single person or instrument on the same item, under the same conditions, and in a short period of time. It is also known as test–retest reliability. Variation in measurements under these conditions is known as equipment variation (EV).

Reproducibility (appraiser variation) is the variation in the averages of measurements made by different operators using the same measuring instrument when measuring one part. Variation under these conditions is known as appraiser variation (AV).

It is affected due to factors other than machine variation. These factors include operators, different instruments, temperature, humidity, and part fixing or holding methods.

Reproducibility is a limit within which agreement may be expected 95% of the time between two test results obtained in different measurement laboratories for the same compatible sample of material.

MSA can separate reproducibility into its component parts and can be used to decide which component needs to be improved.

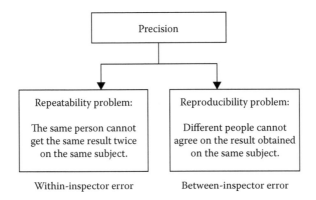

FIGURE 5.8
Two components of precision.

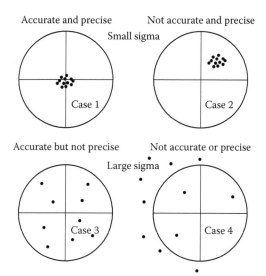

FIGURE 5.9
Measurement system errors.

According to AIAG (2002), a general rule of thumb for measurement system acceptability is (Figure 5.9)

- Variation or error under 10% is acceptable.
- Variation or error between 10% and 30% suggests that the system is acceptable, depending on the importance of the application, cost of the measurement device, cost of the repair, and other factors.
- Variation or error more than 30% is considered unacceptable, and you should improve the measurement system.

Repeatability and reproducibility are related as follows:

Gage variance = Repeatability variance + Reproducibility variance

$$\sigma^2 \text{ Gage} = \sigma^2 \text{ Repeatability} + \sigma^2 \text{ Reproducibility}$$

Measurement System Process

A measurement system can have any of the above problems.

Case 1: The measurement instrument measures parts with very small variation and is very accurate—all measurements are very close to the actual.

Case 2: The measurement instrument measures parts with very small variation but is not accurate.

Case 3: We could also have an instrument where the mean of the measurements is very close to the actual value, but the measurement data have a large variance (not precise).

Case 4: Finally, there could be a measurement instrument that is neither accurate nor precise.

The MSA process steps are as follows:

1. Prepare for the study.
2. Understand and evaluate stability.
3. Evaluate resolution.
4. Determine accuracy.
5. Calibrate the instrument.
6. Evaluate linearity
7. Determine repeatability and reproducibility.

Step 1: Prepare for the Study

Objective

Determine which process parameters will be used for the rest of the study.

Process

- Determine what measurement system is to be studied. This includes the instrument and the test procedure.
- Document the test procedure. Train the operators in understanding the procedure.
- Determine the number of operators (k), the number of sample parts (n), and the number of repeated readings (r). As a rule of thumb, you may choose 3 operators, 4–10 sample parts, and 3–6 repeated readings.
- Note: The greater the number of operators or parts or readings, the greater the confidence and the greater the price—so you decide how the confidence should be utilized.
- Determine who the operators will be. The operators chosen should be selected from those who are trained or certified on the measurement system.

- Determine what samples will be used and where the samples will come from. The sample parts must come from an existing process and represent its entire operating range.
- Several methods can be used, which will ensure randomly distributed sample parts. The most common method includes taking samples on different days during different shifts.
- The instrument should have a resolution that is no more than 1/10 of the total process variation. For example, if the variation within the characteristic is 0.05, the instrument should read a change of 0.005 or less.

Step 2: Understand and Evaluate Stability

- Stability is the change in bias over time.
- Stability is the consistency of performance over time.
- Stability can be evaluated using a control chart and plotting measurements on the same part over time.

Objective

Determine if the measurement system is stable (in statistical control) over time.

Process

- Choose either
  1. One sample standard. If one is not available, use a production unit whose measurement falls into the middle of the expected measurement range.
  2. Several sample standards from the low end, middle, and high end of the expected measurement range. If they are not available, use production units. If several samples are chosen, separate control charts and data analyses will need to be performed.
- Measure the sample standard three to five times in as short a period of time as possible to obtain one group of data. This should be performed on a regular basis, such as hourly, daily, or weekly.
- Initially, 25 groups of data should be gathered.
- Plot the data on an average (X) control chart, pronounced $\bar{X}$, and a range (R) control chart with their appropriate control limits.

  $\bar{X}$ and R charts are commonly used because the $\bar{X}$ evaluates the process's central tendency over time, while the range

(R) evaluates the process spread over time. Calculating $\bar{X}$ and R requires the steps presented below.

$\bar{X}$ and R Chart Example

Step 1: The means and ranges of each sample are calculated first (Figure 5.10; UCL = upper control limit; LCL = lower control limit).

Step 2: For $n = 5$, find A2, D4, and D3 (Table 5.3).

Calculate the necessary table values and enter the results as shown in Figure 5.11.

Prepare graphs (Figure 5.12).

The advantages of the $\bar{X}$ and R chart are

- They are easy to construct.
- They are easy to interpret.
- They are used when a process can be sufficiently monitored by collecting variable data in small subgroups.
- They can be sensitive to process changes and provide early warning and opportunity to act before a situation worsens.

The disadvantage is that the $\bar{X}$ and R chart can be used only when data are available to collect in subgroups.

Evaluation of the control charts for unstable and out-of-control conditions is shown in Figure 5.13.

Title								
	Aug 1	Aug 2	Aug 3	Aug 4	Aug 5	Aug 6	Aug 7	Aug 8
1	25.3	24.4	25.3	25.0	25.3	24.9	25.6	25.3
2	25.2	24.9	24.9	24.8	25.5	24.9	26.0	24.7
3	24.9	24.8	26.1	27.5	27.2	25.1	25.3	26.4
4	26.4	25.5	25.6	26.0	25.5	25.3	25.3	25.0
5	25.0	25.2	25.3	25.4	26.4	26.5	25.4	25.8
Mean								
Range								
Grand mean $\bar{\bar{X}}$				Range mean $\bar{\bar{R}}$				
UCL $_{\bar{X}}$				UCL $_{\bar{R}}$				
LCL $_{\bar{X}}$				LCL $_{\bar{R}}$				

FIGURE 5.10
Data for stability test.

TABLE 5.3

Factors Used in $\bar{X}$ and R Study

	Control Chart Constants							
	Chart for X		Chart for Std. Deviation			Chart for Ranges		
Sample Observations	Control Limit Factors		Centerline Factors	Control Limit Factors		Centerline Factors	Control Limit Factors	
n	A_2	A_3	C_4	B_3	B_4	d_2	D_3	D_4
2	1.880	2.659	0.7979	0	3.267	1.128	0	3.267
3	1.023	1.954	0.8862	0	2.568	1.693	0	2.574
4	0.729	1.628	0.9213	0	2.266	2.059	0	2.282
5	0.577	1.427	0.9400	0	2.089	2.326	0	2.114
6	0.483	1.287	0.9515	0.030	1.970	2.534	0	2.004
7	0.419	1.182	0.9594	0.118	1.882	2.704	0.076	1.924
8	0.373	1.099	0.9650	0.185	1.815	2.847	0.136	1.864
9	0.337	1.032	0.9693	0.239	1.761	2.970	0.184	1.816
10	0.308	0.975	0.9727	0.284	1.716	3.078	0.223	1.777
15	0.223	0.789	0.9823	0.428	1.572	3.472	0.347	1.653
20	0.180	0.680	0.9869	0.510	1.490	3.735	0.415	1.585
25	0.153	0.606	0.9896	0.565	1.435	3.931	0.459	1.541

Figure 5.13 defines out-of-control conditions as

- A single point outside the control limits. In the figure, point 16 is above the UCL.
- Two out of three successive points are on the same side of the centerline and farther than 2 sigma from it. In the figure, points 3 and 4 send that signal.
- Four out of five successive points are on the same side of the centerline and farther than 1 sigma from it. In the figure, points 7–11 send that signal.
- A run of eight in a row are on the same side of the center-line—or a run of 10 out of 11, 12 out of 14, or 16 out of 20. In the figure, point 21 is eighth in the row above the centerline.
- Other obvious patterns suggest there is something unusual about the data and the process.

Analysis

Both the $\bar{X}$ and R control charts should appear to satisfy the above conditions. If either of them is not stable, a root cause needs to

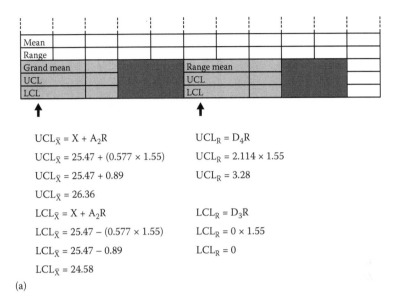

$UCL_{\bar{X}} = X + A_2R$ $UCL_R = D_4R$

$UCL_{\bar{X}} = 25.47 + (0.577 \times 1.55)$ $UCL_R = 2.114 \times 1.55$

$UCL_{\bar{X}} = 25.47 + 0.89$ $UCL_R = 3.28$

$UCL_{\bar{X}} = 26.36$

$LCL_{\bar{X}} = X + A_2R$ $LCL_R = D_3R$

$LCL_{\bar{X}} = 25.47 - (0.577 \times 1.55)$ $LCL_R = 0 \times 1.55$

$LCL_{\bar{X}} = 25.47 - 0.89$ $LCL_R = 0$

$LCL_{\bar{X}} = 24.58$

(a)

Title								
	Aug 1	Aug 2	Aug 3	Aug 4	Aug 5	Aug 6	Aug 7	Aug 8
1	25.3	24.4	25.3	25.0	25.3	24.9	25.6	25.3
2	25.2	24.9	24.9	24.8	25.5	24.9	26.0	24.7
3	24.9	24.8	26.1	27.5	27.2	25.1	25.3	26.4
4	26.4	25.5	25.6	26.0	25.5	25.3	25.3	25.0
5	25.0	25.2	25.3	25.4	26.4	26.5	25.4	25.8
Mean	25.36	24.96	25.44	25.74	25.98	25.34	25.52	25.44
Range	1.5	1.1	1.2	2.7	1.9	1.6	0.7	1.7
Grand mean $\bar{\bar{X}}$	25.47			Range mean $\bar{R}$	1.55			
$UCL_{\bar{X}}$	26.36			$UCL_{\bar{R}}$	3.28			
$LCL_{\bar{X}}$	24.58			$LCL_{\bar{R}}$	0			

(b)

FIGURE 5.11

Calculated UCL and LCL values.

be found and fixed. To validate the fix, take more data to prove that the control conditions are satisfied.

An out-of-control R chart suggests that the process variability within a subgroup is unstable. This may happen if the parts were moving in the fixture. An out-of-control X chart suggests that some

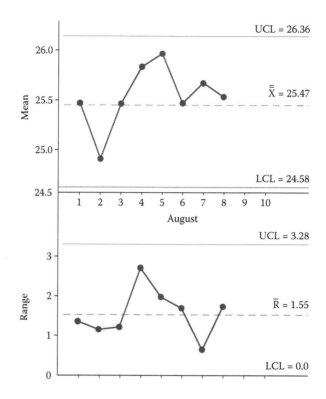

FIGURE 5.12
Mean and range charts.

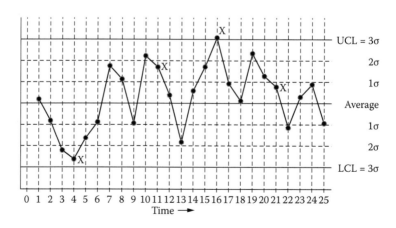

FIGURE 5.13
Chart showing out-of-control conditions.

factor has caused a shift in the measurements. This can be from a change in environment, improper fixing, or a change in instrument.

Overadjustment
- Overadjustment is the result of "tinkering" with the process, or adjusting to swings in the process that are really the result of common-cause variation.
- The rules for detecting the presence of special causes are designed to prevent tinkering and make sure that when the control chart gives a signal, it can be investigated to make the process stable and efficient.

Underadjustment

Underadjustment occurs due to ignoring an out-of-control signal or not monitoring the control chart. Underadjustment results in the loss of opportunities for improvement.

Maintaining Control

It is important to maintain control in the measurement system. If the measurement system is out of control, any measurements made with it will be uncertain and not represent the true value of the characteristic being measured.

Every time the measurement system is prepared for measuring a characteristic, a set of measurements should be made on the same sample standard as above. The data should then be plotted on the same control chart as above. The control chart should be evaluated for any unstable conditions.

The measurement system should be checked for its stability for each measurement project.

Step 3: Evaluate Resolution

Objective

Determine if the measurement system can detect and discriminate between small changes in the measured characteristic.

Note: This process can be performed concurrently with the previous process, evaluation of stability.

Process
- Choose a sample standard. If one is not available, use a production unit whose measurement falls into the middle of the expected measurement range.

- Measure the sample standard three to five times in as short a period of time as possible to obtain one group of data.
- Repeat the measurement process to obtain a minimum of 10, and preferably 25, groups of data. Each of the repeated measurement sets should be performed on a regular basis, such as hourly, daily, or weekly.
- Plot the data on an R control chart with its appropriate control limits.

Analysis

If any of the following exist, then the resolution is inadequate:

- There are only one, two, or three possible values for the range.
- There are only four possible values for the range when the subgroup size is greater than or equal to 3 ($n \geq 3$).

Step 4: Determine Accuracy

Objective

Determine the closeness of agreement between an experimental measurement of a characteristic and the true value of that characteristic.

Process

- Choose either
  1. One sample standard from the middle of the expected measurement range.
  2. Several sample standards from the low end, middle, and high end of the expected measurement range. If sample standards are not available, use production units from the expected measurement range.
- Have the samples measured 15 times on a measurement system with a known bias (calibration lab).
- Compute the average of the 15 readings. Use this average as the reference value.

Note: Measure the sample standard 15–25 times in as short a period of time as possible using the same operator, same equipment, and same setup to obtain the data.

- Compute the average of the readings, $\bar{X}$ and S.
- Compute the bias using the following equation:

$$\text{Bias} = \text{Average} - \text{Reference value}$$

Step 5: Calibrate the Instrument

Objective

Align the measurement instrument so that its measurement bias is minimized when compared to some traceable standard.

Process

Calibrate the instrument according to the manufacturer's instructions. This should be accomplished prior to performing any measurements and should be performed by an independent source, such as a metrology laboratory.

Step 6: Evaluate Linearity

Linearity is the difference in the bias (Y) values through the expected operating range of the gage reference value (X).

The linearity principle is to establish a statistical linear relationship between two sets of corresponding data (X and Y) by fitting the data to a straight line by means of the least-squares technique or regression analysis.

Linearity can be calculated as follows:

Linearity = Slope × Process variation or through a coefficient of regression

The resulting linearity line takes the general form

$$y = bx + a \tag{5.2}$$

where a is the intercept of the line with the y-axis and b is the slope (tangent).

In regression analysis, one factor has to be independent. This factor is by convention designated as X, whereas the other factor is the dependent factor, Y (thus, we speak of regression of Y on X).

The line parameters b and a are calculated with the following equations:

$$b = \frac{\sum (X_i - \bar{X})(Y_i - \bar{Y})}{\sum (X_i - \bar{X})^2}$$

and

$$a = \bar{y} - b\bar{x} \tag{5.3}$$

The correlation coefficient r can be calculated by

$$r = \frac{\sum (X_i - \bar{X})(Y_i - \bar{Y})}{\sqrt{\sum (X_i - \bar{X})^2 \sum (Y_i - \bar{Y})^2}} \tag{5.4}$$

Table 5.4 shows an example.
Using the above data and the formulas, we have

$$b = 0.037$$

$$a = 0.626$$

The regression line equation is

$$Y = 0.626X + 0.037$$

and the regression coefficient is

$$r = 0.997$$

TABLE 5.4

Linearity Data

x_i	y_i	$x_1 - \bar{x}$	$(x_i - \bar{x})^2$	$y_i - \bar{y}$	$(y_i - \bar{y})^2$	$(x_i\bar{x})(y_i\bar{y})$
0	0.05	−0.5	0.25	−0.3	0.09	0.15
0.2	0.14	−0.3	0.09	−0.21	0.044	0.063
0.4	0.29	−0.1	0.01	−0.06	0.004	0.006
0.6	0.43	0.1	0.01	0.08	0.006	0.008
0.8	0.52	0.3	0.09	0.17	0.029	0.051
1	0.67	0.5	0.25	0.32	0.102	0.16
3	2.1	0	0.7	0	0.2754	0.438 Σ
$\bar{x} = 0.5$	$\bar{y} = 0.35$					

Note: x_i = Gage reference value, in our case, observed concentration value; $\bar{x}$ = mean value of x; y_i = bias or, in our case, absorbance; $\bar{y}$ = mean value of y.

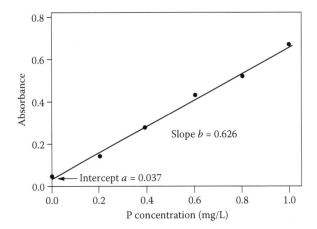

FIGURE 5.14
Linearity graph.

The graph is shown in Figure 5.14.

Coefficient $r = 0.997$ is very close to 1, and therefore the relation of x and y is said to be linear.

However, if the value is not close to 1, say it is below 0.975, the linearity is in question and it is advisable to repeat the procedure.

The correlation coefficient only gives an indication about how well the best-fit line accounts for the variability in the x–y graph. It does not tell us how well the measurement instrument performs.

Step 7: Determine Repeatability and Reproducibility

Objective

Determine the amount of variability in a set of measurements taken on a single measurement instrument that can be attributed to the measurement instrument itself (repeatability) and to the entire measurement system (reproducibility).

Process

Completely randomize the samples (n) and operators (k) to determine what the measurement order will be for the first run.

For example, if there are three samples and two operators, one possible random order might be as shown in Table 5.5.

Repeat the randomization process for each of the subsequent runs.

TABLE 5.5

Example of Measurement Order

Measurement Order	$k = 2$ Appraiser	$n = 3$ Sample
1	B	1
2	B	3
3	A	2
4	B	2
5	A	3
6	A	1

1. Have the operators measure the samples in the measurement order that was just determined.
2. The engineer should read and record the data so that the data value remains hidden from the operator. Try to ensure that the operators do not know each other's readings.
3. The data can then be entered into a software program such as MINITAB, where it can be analyzed using the "Gage R&R Study—ANOVA method."

Analysis

- Plot the data (see Figure 5.17).
- Perform an analysis of variance (ANOVA) on the data. The potential sources of variability can be designated as follows:
  Operator variance σ_O^2
  Parts (product) variance σ_P^2
  Operators by parts σ_{OP}^2
  Repeatability $\sigma_{Repeatability}^2$
- Calculate the total variation, $Z_O^* \sigma_{Total}$:

$$\sigma_{Total} = \sqrt{\sigma_P^2 + \sigma_{Repeatability}^2 + \sigma_{Reproducibility}^2} \qquad (5.5)$$

where

$$\sigma_{Reproducibility}^2 = \sigma_{Operator}^2 + \sigma_{OP}^2$$

Z_O in the equation is used to assume a certain percent confidence in the spread of the data. A typical value of 5.15 is used to ensure a 99% confidence. See Table 5.6.

TABLE 5.6

Values for Z_O at a Given Confidence Level

% Confidence	Z_O
90	3.29
95	3.92
99	5.15
99.73	6

- Calculate the percent contribution of each of the components of variation using the following equation:

$$\% \text{ Contribution} = \left(\frac{\sigma^2_{\text{Component}}}{\sigma^2_{\text{Total}}} \right) * 100\% \qquad (5.6)$$

where $\sigma_{R\&R} = \sqrt{\sigma^2_{\text{Repeatability}} + \sigma^2_{\text{Reproducibility}}}$.

- Determine if the percent contribution (R&R) is acceptable using the following guideline: the percent contribution (R&R) should be less than 30%. If the percent contribution (R&R) is greater than 30%, work should be refocused on reducing measurement variability before any further analysis is performed. If there is a specified tolerance for the characteristic being measured, use the following four steps to determine the process-to-tolerance ratio (P/T):

  1. Calculate the P/T for repeatability using the following equation:

$$P/T_{\text{Repeatability}} = \left(\frac{Z_O * \sigma_{\text{Repeatability}}}{\text{USL} - \text{LSL}} \right) * 100\% \qquad (5.7)$$

  2. Determine if the P/T (repeatability) is acceptable using the following guideline: If P/T (repeatability) is small (i.e., <5%), then it is acceptable. This will help to ensure attainment of acceptance of percent contribution (R&R). If the P/T (repeatability) is greater than 5%, then the measurement system may not be adequate for this application.

  3. Calculate the P/T for the measurement instrument using the following equation:

$$P/T = \left(\frac{Z_0 * \sigma_{R\&R}}{USL - LSL} \right) * 100\% \qquad (5.8)$$

where $\sigma_{R\&R} = \sqrt{\sigma_{Repeatability}^2 + \sigma_{Reproducibility}^2}$.

4. Determine if the measurement instrument P/T is acceptable using the following guideline: percent contribution (R&R) should be less than 30%. If the percent contribution (R&R) is greater than 30%, work should be refocused on reducing measurement variability before any further analysis is performed.

In reading the equipment, the readings should be estimated to the nearest number that can be obtained. If possible, readings should be made to one-half of the smallest graduation. For instance, if the smallest graduation is 0.0001, then the estimate for each reading should be rounded to the nearest 0.00005.

The study should be observed by a person who recognizes the importance of the caution required in conducting a reliable study.

The measurement procedure should be documented and all operators trained for the procedure prior to the study.

Each operator should use the same procedure, including all steps to obtain the readings:

Step 1: Choose 10 samples and number them. Select two capable appraisers (inspectors).

Step 2: Calibrate the gage or use a calibrated gage.

Step 3: Have the first inspector measure all samples in a random order.

Step 4: Have the second inspector measure all samples in a random order.

Step 5: Continue until all operators have measured the samples once (trial 1).

Step 6: Repeat steps 3–5, making sure that the inspectors do not know previous results.

Fill in the repeatability and reproducibility data sheet (Figure 5.15).

	Operator A			Operator B			Operator C		
	Measurements			Measurements			Measurements		
Sample	1	2	3	1	2	3	1	2	3
1									
2									
3									
4									
5									
6									
7									
8									
9									
10									
Notes:									

Gage name: _____ Date: _____ Performed by: _____
Gage no: _____ Dimension min: _____ Dimension max: _____
Part name: _____ Characteristic: _____

FIGURE 5.15
Repeatability and reproducibility data sheet.

Step 7: Use the constants and formulas of Figure 5.16 and calculate the ranges, average range, UCL, and LCL as explained in the repeatability and reproducibility report (Figure 5.17).

Step 8: Calculate repeatability (equipment variation). This is a variation in measurements under exact conditions known as equipment variation (EV), due solely to gages.

To ensure 99% confidence, a value of 5.15 sigma spread is used. See below for values of sigma for a given confidence limit:

% Confidence	Sigma (Z)
90	3.29
95	3.92
99	5.15
99.73	6

$$EV = \bar{R} * K1 = 9.5 \times 4.56 = 43.3$$
$$\sigma EV = \bar{R}/d2 = 9.5/1.13 = 8.4 \text{ OR} \qquad (5.9)$$
$$\sigma EV = EV/5.15 = 43.3/5.15 = 8.4$$

Repeatability—equipment variation (EV)	$EV = \overline{\overline{WR}}{}^{*}K_1$	$\overline{\overline{WR}}$ = Average of the within range of the trials of each part (R_e)		
	$\%EV = 100\,[EV/TV]$	Trials (r)	2	3
		K_1	4.56	3.05
Reproducibility—appraiser variation (AV)	$AV = \sqrt{(\overline{X}_{Diff} * K_2)^2 - \dfrac{(E.V.)^2}{n^*r}}$	$\overline{X}_{Diff}$ = Range of the operator averages (R_o)		
	$\%AV = 100\,[AV/TV]$	Operators (k)	2	3
	n = number of parts	K_2	3.65	2.70
	r = number of trials			
Part variation (PV)	$PV = R_p{}^{*}K_3$	R_p = Range of the part averages		
	$\%PV = 100\,[PV/TV]$	Parts (n)	5	10
		K_3	2.08	1.62
Repeatability and reproducibility (R&R)	$R\&R = \sqrt{(EV)^2 + (AV)^2}$	%R&R<10%—process is capable		
	$\%R\&R = 100\,[R\&R/TV]$	10%<%R&R<20%—barely capable		
		more than 20%—requires improvement		
Total variation (TV)	$TV = \sqrt{(R\&R)^2 + (PV)^2}$			

FIGURE 5.16

Summary of equations for one standard of deviation used in the gage R&R study. (From Ford Motor Company, Fiat Chrysler Automobiles [FCA], and General Motors [GM] Measurement System Analysis [MSA], 3rd ed., March 2002, pp. 118–119.)

Step 9: Calculate the reproducibility (appraiser variation). This is a variation in the average of measurements, when different operators measure the same part. It is known as appraiser variation (AV).

Figure 5.17, the repeatability and reproducibility report, shows

$$AV = 39.7$$

$$\sigma AV = 7.7$$

Precision can be separated into two components, called repeatability and reproducibility, which are related as follows:
Gage variance = Repeatability variance + Reproducibility variance
$\sigma 2$ Gage R&R = $\sigma 2$ Repeatability + $\sigma 2$ Reproducibility
σ R&R = 11.4
% Repeatability = 16.8%
% Reproducibility = 15.4%
% R&R = 22.8%

Repeatability and Reproducibility Report

Product Characteristic	Aluminum analysis	Date
Upper Specification Limit	650	Performed By
Lower Specification Limit	350	Gage
Process Sigma	50	Gage Number
Number of Operators	2	

Oper.	A				B				C				D			
Sample #	1st Trial	2nd Trial	3rd Trial	Range	1st Trial	2nd Trial	3rd Trial	Range	1st Trial	2nd Trial	3rd Trial	Range	1st Trial	2nd Trial	3rd Trial	Range
1	471	484		13	485	480		5								
2	766	742		23	778	807		29								
3	328	326		2	328	314		14								
4	446	433		13	455	450		5								
5	456	454		2	470	461		9								
6	443	450		7	460	455		5								
7	552	557		5	551	547		4								
8	477	479		2	499	492		7								
9	509	508		1	513	540		27								
10	384	371		13	390	385		5								
Totals	4831	4804		81	4929	4931		110								
				8.1				11								

Sum$_A$	9635	$\bar{R}_A$	Sum$_B$	9860	$\bar{R}_B$	Sum$_C$	$\bar{R}_C$	Sum$_D$	$\bar{R}_D$
$\bar{X}_A$	481.8		$\bar{X}_B$	493.0		$\bar{X}_C$		$\bar{X}_D$	

$\bar{R}_A$	8.1
$\bar{R}_B$	11
$\bar{R}_C$	
$\bar{R}_D$	
Sum	19.1
$\bar{\bar{R}}$	9.5

# Trials	D_4	d_2
2	3.27	1.13
3	2.58	1.69
4	2.28	2.06

$^*UCL_R = D_4 \times \bar{\bar{R}}$

$\bar{R}=$	9.5
$D_4 =$	3.27
$^*UCL_R$	31.1

Maximum $\bar{X}=493.0$
Minimum $\bar{X}=481.8$
$\bar{X}_{DIFF}=11.2$

*Limit of individual Rs. Identify the cause and correct for those beyond the limit. Either a) repeat those readings using the same operator and gage, or b) discard those values and recompute $\bar{R}$ and UCL from the remaining observations.

(a)

Repeatability

Equipment variation (EV) = 99% range

$EV = \bar{R} \times K_1 = 9.5 \times 4.56 = 43.3$

$\sigma_{EV} = \bar{R}/d_2 = EV/5.15 = 8.4$

# Trials	2	3
K_1	4.56	3.05

Reproducibility

Appraiser variation (AV) = 99% range

$$AV = \sqrt{\left(\bar{X}_{diff} \times K_2\right)^2 - \left[\left(EV\right)^2 / (n \times r)\right]} =$$

$$\sqrt{(11.2 \times 3.65)^2 - \left[(43.3)^2 / (10 \times 2)\right]}$$

$A.V. = 39.7 \mid^{**}$ $\sigma_{AV} = \dfrac{AV}{5.15} = 7.7$

n = # of samples
r = # of trials

# of operators	2	3	4
K_2	3.65	2.7	2.3

**If a negative value is calculated under the square root or if there is only one operator, AV = 0.

Repeatability and Reproducibility

R&R = 99% range

$$R\&R = \sqrt{(EV)^2 + (AV)^2} =$$

$$\sqrt{(43.3)^2 + (39.7)^2} = 58.8$$

$\sigma_{R\&R} = \dfrac{R\&R}{5.15} = 11.4$

Measurement Capability☐
Index 1 As a Percentage of☐
Process Variation (σ_t)

% Repeatability $= 100 \times \dfrac{\sigma_{EV}}{\sigma_t}$
$= 100 \times 8.4/50 = 16.8\%$

% Reproducibility $= 100 \times \dfrac{\sigma_{AV}}{\sigma_t}$
$= 100 \times 7.7/50 = 15.4\%$

% R&R $= 100 \times \dfrac{\sigma_{R\&R}}{\sigma_t}$
$= 100 \times 11.4/50 = 22.8\%$

% R&R Acceptance Criteria MCI-1	
Good	>0 – <20%
Marginal	>20%; <30%
Unacceptable	>30%

(b)

FIGURE 5.17
Repeatability and reproducibility report.

% R&R acceptance criteria are as follows:

Good	>0 to <20%
Marginal	>20% to <30%
Unacceptable	>30%

See Figure 5.16 for the AIAG recommended summary of equations for the gage R&R study.

In the worksheet in Figure 5.17, 2 appraisers and 10 parts are selected. It uses the formulas per the Figure 5.16 above, as well as the values of K constant.

PROCESS CAPABILITY MEASUREMENT

We have studied measurement system analysis and made sure the measurement method has been analyzed and has an acceptable level of measurement error. Now we are ready to measure the capability of a process. Process needs to stay in control within the specifications (called process control limits) required by the customer.

Process limits are the voice of the process resulting from the product variations produced. The supplier collects data over time to form a process curve, known as a normal curve, for determining the variation in the units against the customer's specification.

A histogram is a special type of graph that allows you to see the variation in your process. It is sometimes called a frequency distribution.

In histograms, frequency is represented by the area of each column, unlike many bar charts, where the height alone of each column represents the frequency.

Consider a simple example where an army recruitment center wants to recruit soldiers with a target height of 69 inches out of 50 selected candidates. The height data that were collected from a group of 50 candidates are shown in Table 5.7.

There are only 50 measurements, but it is difficult to draw specific conclusions about the data without further analysis. A histogram can be constructed to provide more usable information.

Arrange the data as shown in Table 5.8.

The histogram will look like that shown in Figure 5.18.

TABLE 5.7

Individual Heights Measured in Inches

69.9	68.9	68.2	66	71
69	70	68.5	66.5	72.5
69.6	69.5	70	67.5	73
68.5	70.4	66.8	68.3	69
65	71.1	69	68.2	71.3
65.9	71	69.3	69.1	68.2
67.2	72.5	69.1	70.2	68.5
67.5	73.1	69.4	69.5	70
68	68.8	68.5	70.5	67
68.6	71.3	65.5	70.8	69.2

Note: Data average is 69.125 inches. Sigma = 7.5.

TABLE 5.8

Height Data Frequency Diagram

Class	Height Intervals	Frequency	Total
1	64.4–65.0	X	1
2	65.1–65.6	X	1
3	65.7–66.2	XX	2
4	66.3–66.8	XX	2
5	66.9–67.4	XX	2
6	67.5–68.0	XXXX	4
7	68.1–68.6	XXXXXXXXX	9
8	68.7–69.2	XXXXXXXXX	9
9	69.3–69.8	XXXXXXX	7
10	69.9–70.4	XXX	3
11	70.5–71.0	XXXX	4
12	71.1–71.6	XX	2
13	71.7–72.2	XX	2
14	72.3–72.8	XX	2

The following is found:

Mean = 69.125
Sigma (s) = 1.87 (s stands for short term sigma)
Spread = 7.5 = 2.2 s
$3\,s$ = 3.61
$6\,s$ = 11.22

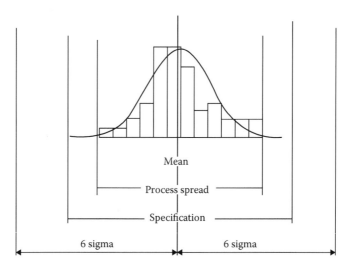

FIGURE 5.18
Basic process capability diagram.

USL = 70
LSL = 60
Cp = (70−60)/11.22 = 0.891

Once the histogram is developed, you can analyze the data with regard to customer expectations (specifications). You can see from Figure 5.19 that the first histogram of a process sample falls within the specifications, while the second has a portion of the histogram outside of the specifications.

The second histogram has too much dispersion, or variability, to meet customer expectations. The indication is that action must be taken to make the output more consistent with the customer requirement, or some number of defects (candidates outside the required specification) will be produced.

The ideal process remains within the specifications, producing 100% conforming products, and is predictable, as shown in the upper histogram.

Process Capability Assessment

Cp, Cpk, Pp, Ppk

- These are nondimensional constants used to describe capability.
- In Six Sigma organizations, they are more useful than percentage yields.

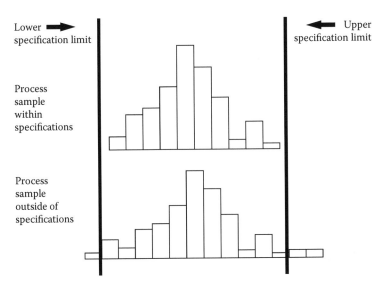

FIGURE 5.19
Process capability histograms.

- Most capability index estimates are valid only if the sample size used is "large enough." Large enough is generally thought to be about 50 independent data values.
- The Cp, Cpk, and Cpm statistics assume that the population of data values is normally distributed. Assuming a two-sided specification, if $\bar{X}$ and S are the mean and standard deviation, respectively, of the normal data and USL, LSL, and T are the upper and lower specification limits and the target value, respectively, then the population capability indices are defined as follows:

Cp

$$Cp = \frac{USL - LSL}{6 * S_{\text{short-term}}} \qquad (5.10)$$

(USL − LSL) is called the tolerance band or the specification band. If the tolerance band is equal to 1 sigma on either side of the mean (2 sigma total), the sigma level is 1,

$$Cp = 2/6*1 = 1/3$$

and the process is said to be at 1 sigma.

If the spread is equal to 3 sigma on either side of the mean (6 sigma total),

$$Cp = 6/6*1 = 1$$

and the process is said to be at 3 sigma.

If this spread is equal to 6 sigma on either side of the mean (12 sigma total),

$$Cp = 12/6*1 = 2$$

and the process is said to be at 6 sigma.

In short, Cp = Sigma level/3.

Cpk

- The difference between Cp and Cpk is that Cp assumes the voice of the process is centered halfway between the sigma limits and Cpk uses the actual voice of the process mean.
- The bigger the difference between Cp and Cpk, the greater opportunity there is to improve the process capability by centering. For some simple processes, this is valuable information as you only need change an offset to increase capability:

$$Cpk = \min\left(Cpk(USL), Cpk(LSL)\right)$$
$$Cpk(USL) = \frac{(USL) - \bar{X}}{3 * s_{short-term}}$$
$$Cpk(LSL) = \frac{(\bar{X} - LSL)}{3 * s_{short-term}} \qquad (5.11)$$

For example, a tea bag packing machine packs the tea between 39 and 49 grams in each bag. The packing process has a process standard deviation of 2 grams, and the mean weight of the tea in bags is 42 grams. What is the process capability of the process?

$$Cpk \ max = (49 - 42)/3 * 2 = 1.167$$
$$Cpk \ min = (42 - 39)/3 * 2 = 0.50$$
$$Answer: Cpk = 0.5$$

Pp and PPk

The main difference between the Pp and Cp studies is that during the evaluation of Cp, the samples are produced at the same time and the process contains mainly the causes of variation within the sample groups. The standard deviation is therefore lower in the case of the evaluation of Cp.

During the evaluation of Pp, variation between sample groups is added, which enhances the *s* value, resulting in more conservative Pp estimates. The inclusion of between-group variation in the calculation of Pp makes the result more conservative than the estimate of Cp.

The formulas for Pp and Ppk are the same as those for Cp and Cpk.

- The only difference between calculating Cp and Cpk, and Pp and Ppk is that you use the short-term sigma level for C and the long-term sigma level for P.
- Remember, long-term capability = short-term capability + 1.5.
- Also remember that over a long period we expect the capability of a process to deteriorate.
- So referring to Table 5.9, if a project team achieves a process that is 99% capable (short term), we expect it to be 80% capable (long term); and to create a process that is 50% good (long term), we aim for 93% (short term).
- As a guide,
  - Long term is more likely to include special causes.
  - Long term is likely to include mixtures of batches and parts and changing personnel.

TABLE 5.9

Short- and Long-Term Sigma

Short-Term Sigma	Long-Term Sigma	Yield	Long-Term Yield
6.0	4.5	99.9997%	99.865%
5.5	4	99.9968%	99.379%
5.0	3.5	99.9767%	97.725%
4.5	3	99.865	93.319%
4.0	2.5	99.3790%	84%
3.5	2	97.72%	69%
3.0	1.5	93%	50%
2.5	1	84%	31%
2.0	0.5	69%	16%
1.5	0	50%	7%

Cpm

When should we use Cpm?

- The target is not the center or mean of the USL–LSL.
- We need to establish an initial process capability during the measure phase.

The following information is needed:

- Value of the target
- Continuous data
- Sample size number
- Standard deviation (sigma) calculated with actual data, not estimated
- Normal probability distribution knowledge

You are almost never going to use this, but in some processes you will not target the center point.

For example, when cutting impellers, you want to cut an impeller diameter between 196 and 200 millimeters. It may be cheaper to bias the target cut toward, say, 199 millimeters instead of 198 millimeters (you handle optimum fluid for a longer time). So in this case, if we target 199 millimeters, we want to measure the process against the target instead of the central point:

$$Cpm = \frac{Cp}{\sqrt{1+\frac{(\mu-T)}{\sigma^2}}} \tag{5.12}$$

where T is the target value, μ is the expected value, and σ is the standard deviation.

Two Opinions: Process Capability Studies

It may be mentioned here that there are two opinions about process capability studies.

Opinion 1: Process capability describes the overall capability of the process operating at its best. This approach does not address how well the process directly meets customer specifications. The study is usually completed on a short-term basis with a 1.5 sigma adjustment to compensate for drifts in long-term variability.

Opinion 2: Process capability describes how well a process meets customer specifications. This approach takes a longer-term view of variance, and short-term or long-term views are usually not considered separately.

Process Capability Study Procedure

1. Select a process to study.
2. Define and verify the process parameters. Study the process. Prepare a process map to include all process steps and their boundaries.
3. Conduct a measurement system analysis to ensure that the measurement methods produce data with total measurement error well within the limits.
4. Select a process capability analysis method (Cpk, Cp, Ppk, or Pp).
5. Obtain the data and conduct an analysis.
6. Develop an estimate of the process capability. This estimate can be compared to the standards set by internal or external customers.

After completing a process capability study, address any undesirable special causes of variation that can be isolated. In some cases, a special cause of variation may be desirable if it produces a better product or output. In that case, include this special cause in the process so that the product benefits.

MEASURE PHASE SUMMARY

Measure is about measuring what is measurable, that is, data that are to be identified, collected, described, and displayed.

A sound sampling technique ensures data accuracy and integrity. Tools such as Pareto charts and histograms depict the relationship between data.

Measurement system analysis (gage R&R) correlates linearity and precision for assessing the capability of the people involved in the process.

Process capability studies the link between the voice of the customer and the voice of the process. The customer sets the target and specification limits. The provider measures the process results and compares them to the customer's expectations.

Process performance indices such as Cp, Cpk, Pp, and Ppk are numerical values indicating where the process lies in terms of targets, specifications, and sigma levels as per the customer's expectations.

6

Analyze

Analyze the problem. During the analyze phase, potential root causes for the process problem are identified.

The question that the analyze phase seeks to answer is, what is the cause of the problem?

Once the root cause is known, action can be taken in the improve phase to counter it.

The techniques are

- Regression and correlation
- Analysis of variance (ANOVA)
- Hypothesis testing
- Failure mode and effects analysis (FMEA) (see Book 2)
- Additional analysis methods
  - Gap analysis
    - Identify gaps between current performance and goal performance.
    - Prioritize opportunities to improve.
    - Identify sources of variation.
  - Root cause analysis
    - Five whys
    - Pareto chart (see Book 2)
    - Fault tree analysis
    - Cause-and-effect diagram (see Book 2)
  - Waste analysis (nine wastes) (see Book 2)
  - Kaizen (see Book 2)

REGRESSION AND CORRELATION

The statistical data, including the equation to calculate the coefficient of regression, is covered in Chapter 5.

Introduction

In quality improvement projects where the input and output variables are both continuous, quality professionals want to understand if there is a relationship between the two. Correlation and regression are two statistical tools that determine whether a relationship exists between two groups (Figure 6.1). If related, then the question revolves around the relationship's strength.

Correlation measures the strength of the relationship and implies that a change in one variable also changes the other. On the other hand, regression provides a prediction model for the relationship and determines the significance of the factors involved.

Variables and Relationships

Regression is the analysis of relationships between variables. In regression, the relationships of the variables are expressed in the form of Equation 5.3 in Chapter 5. This equation allows for the prediction of the dependent variable (Y) from one or more independent variables (X).

A variable is a quantity able to assume different numerical values. The graph in Figure 6.2 displays the linear relationship between two variables: dependent and independent.

The independent variable X is the input (concentration) that can be set directly to achieve an output result (absorbance).

Before calculating the correlation coefficient, the first step is to construct a scatter diagram. Looking at the scatter diagram will give you a broad understanding of the correlation. Figure 6.3 shows a scatterplot chart example based on an automobile manufacturer.

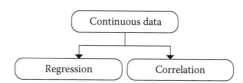

FIGURE 6.1
Continuous data classification.

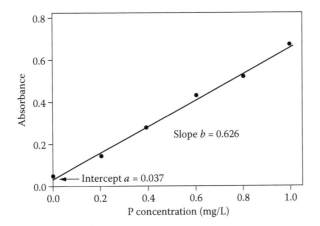

FIGURE 6.2
Graph showing typical relationship between two variables.

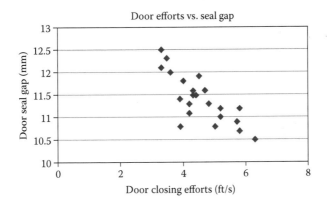

FIGURE 6.3
Correlation between the gap and the door closing effort.

In this case, the process improvement team is analyzing door closing efforts to understand the causes affecting those efforts. The *Y*-axis represents the width of the gap between the sealing flange of a car door and the sealing flange on the body—a measure of how tight the door is set to the body. The fishbone diagram indicates that variability in the seal gap could be a cause of variability in door closing efforts.

In this case, you can see (eyeball) a pattern in the data indicating a negative correlation (negative slope) between the two variables. In fact, the correlation coefficient is 0.77, $[(12 - 11)/(5.3 - 4)]$, indicating a fairly strong relationship.

ANALYSIS OF VARIANCE

ANOVA is a statistical method to determine if more than two population means are equal. The test uses the *F*-distribution (probability distribution) function to compare the amount of variance *within* each population and the amount of variance *between* the population groups to help decide if the variability between and within each population is significantly different.

So the ANOVA test hypothesis can be

$$H_0: \mu_1 = \mu_2 = \mu_3 = \ldots \mu_a$$

Let us understand ANOVA through the following example: Consider that there are three groups (m); m1, m2, and m3. Each group has three members (n); n1, n2, and n3. Now it is proposed that each group undergoes an experiment by consuming three different vitamins: v1, v2, and v3.

To measure the effect of the vitamin consumption on each member, a blood test was carried out.

The results of the test are given in Table 6.1.

Let us first start with calculation of the proportion of variation through a total sum of squares (called SS_T) and then variation within the groups and variation between groups.

Before that, we need to calculate following means:

Mean of the mean of individual group data, also called X-double bar
Means of each group
1 = Mean of group 1 = (3 + 2 + 1)/3 = 2
2 = Mean of group 2 = (5 + 3 + 4)/3 = 4
3 = Mean of group 3 = (5 + 6 + 7)/3 = 6
Grand mean of means = (2 + 4 + 6)/3 = 4

TABLE 6.1

Data to Analyze the Effect of Vitamin Consumption

	v1	v2	v3
	m1	m2	m3
n1	3	5	5
n2	2	3	6
n3	1	4	7

SS_T is defined as the sum of squares of the differences of each group member data point (3, 2, 1, etc.) and the grand mean of means:

$$SS_T = [(3-4)^2 + (2-4)^2 + (1-4)^2] + [(5-4)^2 + (3-4)^2 + (4-4)^2]$$
$$+ [(5-4)^2 + (6-4)^2 + (7-4)^2]$$
$$= 14 + 2 + 14$$
$$= 30$$

SS_T = Total sum of squares = Variation total = 30

Degrees of freedom $(DF)_{Total} = (m \times n) - 1 = (3 \times 3) - 1 = 8$

Now variation within groups (SS_W) is the sum of the squares of the differences of each data point with its group mean. The group means are 2, 4, and 6:

$$SS_W = [(3-2)^2 + (2-2)^2 + (1-2)^2] + [(5-4)^2 + (3-4)^2 + (4-4)^2]$$
$$+ [(5-6)^2 + (6-6)^2 + (7-6)^2]$$
$$= 1 + 0 + 1 + 1 + 1 + 0 + 1 + 0 + 1$$
$$= 6$$

SS_W = Sum of squares within = Variation within groups = 6

Degrees of freedom $(DF)_{Within}$ = Number of independent
data points for each group $(n - 1)$

Multiplied by total number of groups, $m = (n - 1)$, $m = (3 - 1)3 = 6$.
Now variation between groups (SS_B) is the sum of the squares of the differences of each group mean, with the grand mean of the means of individual group data equal to 4:

$$SS_B = [(2-4)^2 + (2-4)^2 + (2-4)^2] + [(4-4)^2 + (4-4)^2 + (4-4)^2]$$
$$+ [(6-4)^2 + (6-4)^2 + (6-4)^2]$$
$$= 12 + 0 + 12$$
$$= 24$$

SS_B = Sum of squares between = Variation between groups = 24

Degrees of freedom $(DF)_B$ = Number of independent groups $(m - 1) = 2$

See Table 6.2 for our conclusion.

TABLE 6.2

Summary of Sums of Squares and Degrees of Freedom

Designation	Sum of Squares	Degrees of Freedom	DF Formula
SS_T	30	8	$mn - 1$
SS_W	6	6	$m(n - 1)$
SS_B	24	2	$m - 1$

Note: $SS_T = SS_W + SS_B$; $DF_T = DF_W + DF_B$; $(mn - 1) = m(n - 1) + m - 1 = (mn - 1)$.

HYPOTHESIS TESTING

Hypothesis testing is a supposition (theory) that is provisionally accepted to interpret certain events or phenomena and to provide guidance for further investigation. A hypothesis may be proven correct or wrong, and must be capable of refutation. If it remains factually undefeated, it is said to be verified or corroborated.

Hypothesis Testing with *F*-Statistics

We found that different vitamin intakes by different groups have different results. We want to find out or infer if this difference is random or if there is really an effect of different vitamins on the groups.

Our null hypothesis is that different vitamins do not make a difference. This means the population means of the three groups are the same:

$$H_0: \mu1 = \mu2 = \mu3$$

An alternative hypothesis is that they are not equal:

$$H_a: \mu1 \neq \mu2 \neq \mu3$$

To test this, we need to use the *F*-statistic:

$$F = (SS_B/DF_B)/(SS_W/DF_W)$$
$$= (24/2)(6/6)$$
$$= 12/1$$

Here we may note that the numerator is much higher than the denominator, and we can therefore infer that there is a difference between populations.

Now let us check this statistically.

Refer to the table at http://mat.iitm.ac.in/home/vetri/public_html/statistics/f-table.pdf. For 10% significance level (90% confidence) and with DF of the numerator horizontally and DF of the denominator vertically, we get the *F*-value as 3.46330.

The calculated *F*-value is much larger than the *F*-statistic 3.46330. So we reject the null hypothesis and conclude that in the experiment, different vitamins did have different effects on these different groups.

Hypothesis Test Definitions

Hypothesis: Premise or claim about a property or a characteristic of a population that we want to test or investigate.

Hypothesis tests: The aim of hypothesis testing is to analyze a sample taken out of a population in an attempt to make a decision about the value of the population characteristic that is likely to occur and the population characteristic that is not likely to occur based on sample data.

Null hypothesis: Denoted by H_0 (H denotes hypothesis and 0 denotes null) and stands for currently accepted value for a population characteristic or population parameter.

Alternative hypothesis: Denoted by H_a (H denotes hypothesis and a denotes alternative). It contains the description of the claim to be tested. The alternative hypothesis is also called a research hypothesis, and this is really what we want to test.

How to Write H_0 and H_a

After a thorough maintenance, a machine fills cans with fluid weighing on average 12 ounces each. But the machine operator thinks that it fills less than 12 ounces.

$$H_0: \mu \geq /= 12 \text{ ounces}$$

$$H_a: \mu \neq 12 \text{ ounces}$$

H_0 and H_a are mathematical opposites.

Here we are testing two means or averages of a given population. We can also test two proportions (percentages) or 2 standard deviations of the given populations.

Two Outcomes of the Hypothesis Test

Reject the null hypothesis H_0 or fail to reject the null hypothesis. How do we test this?

Collect sample data (usually more than 50 data points). In our example, assume that we have weighed 50 cans. Get the average weight and calculate the test statistic. Determination of the test statistic will enable us to draw a line to make a decision.

The level of confidence, C, can be 99%, 98%, or 95%. It tells us how confident we are in our decision.

The level of significance, α, is $1 - C$ if $C = 95\%$:

$$\alpha = 1 - 0.95 = 0.05$$

Let us now carry out a hypothesis test with a sample mean where the variance is known.

Step 1: State the null and alternate hypotheses:

$$H_0: \mu >/= 12 \text{ ounces}$$

$$H_a: \mu \neq 12 \text{ ounces}$$

Choose a sample size of 50 (greater than 30).

Calculate the mean: 11.7, and $\sigma = 1$ standard deviation for the population is the same as the sample because the sample size is large.

Step 2: Choose a significance level α:

$$\alpha = 1 - C \text{ if } C = 95\%$$

$$\alpha = 1 - 0.95 = 0.05$$

Step 3: Calculate test statistic Z:

$$Z = (-X - \mu)/\sigma/\sqrt{n}$$

Calculate Z:

$$Z = (11.7 - 12)/1/\sqrt{50} = -2.12$$

Step 4: Find the p value.

Looking up the normal curve areas for the test statistic –2.12, under Z we find

$$p = 0.5 - 0.4830 = 0.017$$

We find that the p value is less than $\alpha = 0.05$ and reject H_0 and conclude H_a.

The hypothesis test conclusion is that the data provide sufficient evidence to conclude that the mean content of the can is less than 12 ounces.

GAP ANALYSIS

A gap analysis is a tool used to identify a performance difference between a current state and a desired or future state.

The desired or future state may be set by recognizing potential performance determined through such activities as benchmarking or through organizational strategic planning.

Gap analyses are typically performed at the

- Business level: At this level, a business may compare its performance directly to that of competitors or general average industry performance. Gaps at this level are usually financial in nature.
- Process level: At this level, the current performance of a process might be compared to cost, cycle time, and quality characteristics or other processes, or at performance levels required to remain competitive. (Refer to Chapter 4 on quality function deployment.)
- Product level: Gaps at this level are usually identified through differences in features and capabilities, cost, quality, or generally by critical-to-X perspectives.

If the gap between the current state and the future state is sufficiently large, a series of intermediate steps or milestones may be required with more achievable gaps between each one. However, it is important to recognize that gaps are usually never stationary. Therefore, it may be necessary to account for an increasing gap over time, particularly when future states are set relative to competitor performance.

ROOT CAUSE ANALYSIS

Solving a process problem means identifying the root cause and eliminating it. The ultimate test of whether the root cause has been eliminated is the ability to turn the problem on and off by removing and replacing the root cause.

A number of tools for identifying root causes are discussed in Book 2. The five whys and fault tree analysis are two.

Five Whys

Five whys is a technique used to drill down through the layers of causes and effects to the root cause. It consists of looking at an undesired result and asking, why did this occur? When the question is answered, the next question is, why did that occur? and so on. The number five is, of course, arbitrary.

For example, a customer is not satisfied.

Why is the customer not satisfied? The order arrived late.
Why did the order arrive late? It was shipped late.
Why was it shipped late? Final assembly wasn't completed on time.
Why wasn't final assembly completed on time? A part was received from the paint line late.
Why was the part from the paint line late? It arrived late from the fabrication department.
Why was it late from the fabrication department? The fabrication machine was running another part for the same order.
Why was the fabrication machine running another part for the same order? That's the machine always used to run these two parts.
Could they be on separate machines simultaneously when a delivery date is imminent? Yes.

Fault Tree Analysis

Once a failure has been identified as one requiring additional study, a fault tree analysis (FTA) can be performed. Basic symbols used in FTA are borrowed from the electronic and logic fields. The fundamental symbols are the AND and OR gates. Each of these has at least two inputs and a single

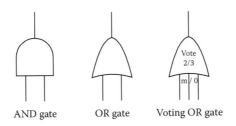

AND gate   OR gate   Voting OR gate

FIGURE 6.4
Common FTA gate symbols.

output. Another key gate is the voting OR gate. In this gate, the output occurs if and only if k or more of the input events occur, where k is specified, usually on the gate symbol. Common FTA gate symbols are depicted in Figure 6.4.

The output for the AND gate occurs if and only if all inputs occur. The output for the OR gate occurs if and only if at least one input occurs. Rectangles are typically used for labeling inputs and outputs. The failure mode being studied is sometimes referred to as the top or head event. An FTA helps the user to consider underlying causes for a failure mode and to study relationships between various failures.

7

Improve

Improve process performance through the following four methods:

1. Prioritization through the cause-and-effect matrix (Pugh matrix)
2. Business process improvement (BPI) and Kaizen
3. Plan–do–check–act (PDCA; focus, fix, and flex)
4. Theory of constraints (TOC)

Then implement the improvements through the following two techniques:

1. Failure mode and effects analysis
2. Process ID (turtle diagram and suppliers, inputs, process, outputs, and customers [SIPOC])

Then evaluate the improvements through the techniques explained in the analysis phase.

Books 1 and 2 cover methods 2 (BPI and Kaizen) and 3 (PDCA).

PRIORITIZATION THROUGH CAUSE-AND-EFFECT MATRIX

1. Identify customer requirements (outputs) from the process map.
2. Assign a priority factor to each output (usually on a 1–10 scale).
3. Identify all process steps and inputs from the process map.

4. Evaluate the correlation of each input to each output.
   • Low score (0, 1, 3): Changes in the input variable (amount, quality, etc.) have a small effect on the output variable.
   • High score (9): Changes in the input variable can greatly affect the output variable.
5. Cross-multiply correlation values with priority factors (step 2) and sum for each input.
6. Rank (sort) the end results.

Figure 7.1 shows how to determine and prioritize paint line process parameters to satisfy customer requirements most economically, expeditiously, and effectively.

Here we can see that curing (score 348) is the most important parameter for the customer.

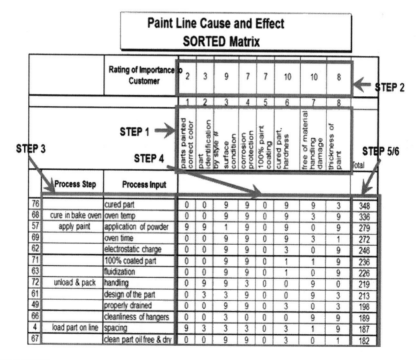

Paint Line Cause and Effect SORTED Matrix

Process Step		Process Input	parts painted correct color	part identification by style #	surface condition	corrosion protection	100% paint coating	cured part, hardness	free of material handling damage	thickness of paint	Total
Rating of Importance Customer			2	3	9	7	7	10	10	8	
			1	2	3	4	5	6	7	8	
76		cured part	0	0	9	9	0	9	9	3	348
68	cure in bake oven	oven temp	0	0	9	9	0	9	3	9	336
57	apply paint	application of powder	9	9	1	9	0	9	0	9	279
69		oven time	0	0	9	9	0	9	3	1	272
62		electrostatic charge	0	0	9	9	0	3	0	9	246
71		100% coated part	0	0	9	9	0	1	1	9	236
63		fluidization	0	0	9	9	0	1	0	9	226
72	unload & pack	handling	0	9	9	3	0	0	9	0	219
61		design of the part	0	3	3	9	0	0	9	3	213
49		properly drained	0	0	9	9	0	3	0	3	198
66		cleanliness of hangers	0	0	3	0	0	0	9	9	189
4	load part on line	spacing	9	3	3	3	0	3	1	9	187
67		clean part oil free & dry	0	0	9	9	0	3	0	1	182

STEP 1, STEP 4 (left side); STEP 2, STEP 3, STEP 5/6 (markers)

FIGURE 7.1
Cause-and-effect matrix.

Improve Process through Lean Six Sigma

Figure 7.2 shows the Lean tools to improve processes.

> **Value stream mapping (VSM):** Define process; identify opportunities for improvements.
>
> **5S:** Instill discipline; free up physical resources.
>
> **Standard work:** Identify value-add; optimize resources with demand; stabilize process.
>
> **Total productive maintenance:** Ensure equipment availability and performance; create capacity.
>
> **Error proofing:** Identify opportunities for errors and eliminate; create capacity.
>
> **Setup reduction:** Create capacity; reduce inventory.
>
> **Continuous flow manufacturing:** Optimize manpower, material, and space; reduce cycle time.
>
> **Pull:** Balance system due to long logistical distances or imbalance in adjacent process cycles.
>
> **Six Sigma:** Reduce process variation; stabilize process for predictability.

THEORY OF CONSTRAINTS

The theory of constraints, created by Dr. Eliyahu Goldratt, is based on the assumption that every system has at least one constraint limiting it from getting more of what it strives for. If this were not true, then the system would produce infinite output. For example, a for-profit organization would make endless profits.

The theory of constraints is both descriptive and prescriptive in nature; it not only describes why system constraints happen, but also offers guidance on what to do about them.

A constraint is anything that limits an organization from moving toward its goal. A constraint can be physical and internal (such as a machine, facility, or policy) or nonphysical and external (such as market conditions or demand for a product).

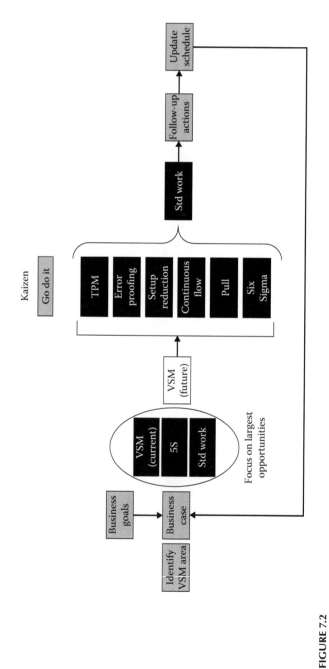

FIGURE 7.2
Lean tools to improve processes.

Five steps for TOC implementation follow:

1. Identify the system constraint and determine whether it is a physical constraint or a policy-related issue.
2. Decide how to exploit the constraint by utilizing every bit of the constraining component without committing to potentially expensive changes or upgrades.
3. Subordinate everything else by adjusting the rest of the system to enable the constraint to operate at maximum effectiveness, and then evaluate the results to see if the constraint is still holding back system performance. If it is, the organization proceeds to step 4. If it is not, the constraint has been eliminated, and the organization skips ahead to step 5.
4. Elevate the constraint. At this point, the organization elevates the constraint by taking whatever action is needed to eliminate it.
5. Go back to step 1, but beware of inertia. After a constraint is broken, the organization repeats the steps all over again, looking for the next thing that constrains system performance. At the same time, it monitors how changes to subsequent constraints may affect the constraints that are already broken, thus preventing solution inertia.

Following are TOC metrics to determine whether an organization is moving toward its goal:

1. Increased throughput (selling price, cost of raw materials)
2. Decreased inventory
3. Decreased operating expenses

The following four measurements are used to identify results for the overall organization:

1. Net profit (NP) = Throughput – Operating expense (T – OE)
2. Return on investment = Net profit/inventory (NP/I)
3. Productivity = Throughput/Operating expense (T/OE)
4. Turnover = Throughput/Inventory (T/I)

8

Control

KEY CONCEPTS AND TOOLS

- Statistical process control (SPC)
- Control chart guide
- Control chart road map
- Control plan
- Sustenance of improvements
  - Lessons learned
  - Implementation of training plan
  - Standard operating procedures (SOPs) and work instructions
  - Ongoing evaluation

STATISTICAL PROCESS CONTROL

SPC Background

SPC was pioneered by Dr. Walter Shewhart in the 1920s and later enhanced by Dr. W. Edwards Deming. SPC is a collection of statistical techniques used to measure and analyze the variation in processes.

Statistical process control (SPC) is not a particular technique or procedure, but a philosophy focusing on optimizing continuous improvement by using statistical tools for analyzing data, making inferences about process behavior, and then making a decision.

> A phenomenon will be said to be controlled, when, through the use of past experience, we can predict, at least within limits, how the phenomenon may be expected to behave in the future.
>
> **Walter Shewhart**

Objectives of SPC

- Produce data to inform and guide process improvement
- Reduce variation
- Increase knowledge about the process
- Quickly detect occurrences of special causes that are shifting the process

Uses of SPC Tools

- Monitor processes for maintaining control
- Detect out-of-control conditions in processes
- Analyze process capability
- Serve as decision-making tools
- Assure customers that product is produced consistently over time
- Enable proactive process improvement

Benefits of SPC

- Assure customers that product is produced consistently over time
- Increase product consistency
- Improve product quality
- Reduce the need for inspection
- Decrease scrap and rework
- Increase production output
- Streamline processes

Basic SPC Concepts

- **Variation:** A change in the process data, a characteristic, or a function that results from some cause
- **Stability:** The condition when observing only random variation around the process target; the absence of special causes of variation; the property of being in statistical control
- **Capability:** The performance of a process demonstrated to be in a state of statistical control
- **Overadjustment:** Making adjustments to a process based on limited data

According to Shewhart, variation in the process is due to two main causes:

1. Common causes of variation are inherent in a process all the time. A process that has only common causes operating is said to be in statistical control. Common cause is sometimes referred to as chance cause, random cause, nonassignable cause, noise, or natural pattern as shown in Figure 8.1 below.

   A process that has only common causes of variation is said to be stable, constant, or predictable and within statistical control over time.

   Examples are

   • Variation in raw material
   • Variation in ambient temperature and humidity
   • Variation in electrical or pneumatic sources
   • Variation within equipment (worn bearings)
   • Variation in the input data
   • Variation in cooking time or temperature
   • Variation in time to clean a hotel room

2. Special causes of variation periodically disrupt the process. A process that has special causes operating is said to lack statistical control. As noted above, Figure 8.2 explains special cause sometimes referred to as assignable cause or unnatural pattern.

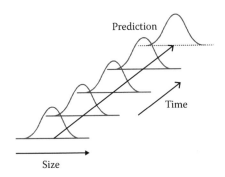

FIGURE 8.1
Common causes.

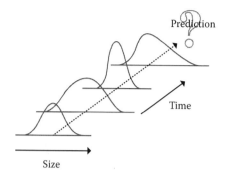

FIGURE 8.2
Special causes.

Special causes of variation have the following characteristics:

- Appear sporadically in a process
- Originate outside the process
- Have greater impact than any single common-cause variation
- Are found in unstable, unpredictable processes

Examples of special causes are

- Excessive tool wear
- Large changes in raw materials
- Broken equipment
- Power reduction (brownout)

Where do you begin? To obtain the greatest benefit from an SPC effort, begin monitoring and improving the most critical operations and processes.

CONTROL CHART GUIDE

- Time is always the horizontal (X) axis.
- Control charts must have
  - Centerline (grand average)
  - Upper control limit (UCL)
  - Lower control limit (LCL)
  - Data points
  - Title

- Legend
- Labeled axes
- Before starting a new control chart, the process must be in control.
- The control limits calculated from the first 20 points are conditional limits.
- Recalculate the control limits after collecting at least 25–30 time-ordered points from a period when the process is operating in control, or whenever a significant and known change to the process has occurred.
- Developing a control chart with fewer than 25–30 points may not be statistically valid.
- If encountering an outlier when developing the control chart, investigate to see if there is an assignable cause; if so, eliminate the point from the analysis.

CONTROL CHART ROAD MAP

The road map in Figure 8.3 is a control chart selection guide. To introduce a particular control chart type later in this chapter, this map will reappear highlighting the path of the particular control chart.

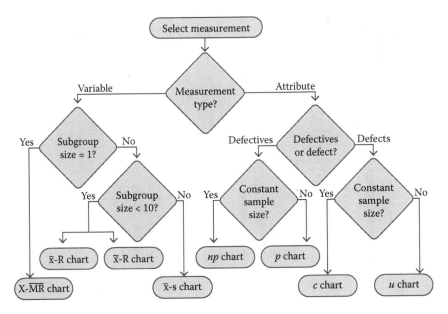

FIGURE 8.3
Control chart selection guide.

Variable Control Charts

Variable control charts monitor key measurable product characteristics or process variables. The formulas assume the normal distribution and limits are established based on ±3 standard deviations; thus, the chance of a part falling outside of the upper or lower control limit is 0.27%.

Variable Data Charts

1. X-bar and R chart (see Chapter 5)
2. X-bar and s chart
3. X-MR chart

Equations for Variable Data Control Charts

We will use the equations mentioned in Figure 8.4 to construct variable data control charts.

Control Chart Constants

Depending on the number of observations in a sample, use of the standard control chart constants as shown in Table 8.1 is made along with the equations of Figure 8.4.

X-Bar and s Chart

Refer to the highlighted position in the road map in Figure 8.5.

X-Bar and s Chart

The formulas are

$\bar{X}$ and s Average and Standard Deviation Chart $\quad \geq 10$

$$\bar{X} = \frac{sum\ of\ subgroup\ averages}{number\ of\ subgroups}$$

$$UCL = \bar{X} + A_3\bar{s}$$

$$LCL = \bar{X} - A_3\bar{s}$$

$$\bar{s} = \frac{sum\ of\ subgroup\ sigmas}{number\ of\ subgroups}$$

$$UCL = B_4\bar{s}$$

$$LCL = B_3\bar{s}$$

Variable Data Charts

CHART TYPE	SAMPLE SIZE	CONTROL LIMITS
$\bar{X}$ and R Average and Range Chart	< 10; usually 3-5	$\bar{X}\ Centerline: \bar{\bar{X}} = \dfrac{(X_1 + X_2 + ...X_k)}{k}$ $UCL = \bar{\bar{X}} + A_2\bar{R}$ $LCL = \bar{\bar{X}} - A_2\bar{R}$ $R\ Centerline: \bar{R} = \dfrac{(R_1 + R_2 + ...R_k)}{k}$ $UCL = D_4\bar{R}$ $LCL = D_3\bar{R}$
$\tilde{X}$ and R Median and Range Chart	<10; odd number of readings in each sample	$\bar{\tilde{X}} = \dfrac{sum\ of\ medians}{number\ of\ medians}$ $UCL_{\tilde{x}} = \bar{\tilde{X}} + \tilde{A}_2\bar{R}$ $LCL_{\tilde{x}} = \bar{\tilde{X}} - \tilde{A}_2\bar{R}$ $R\ Centerline: \bar{R} = \dfrac{(R_1 + R_2 + ...R_k)}{k}$ $UCL = D_4\bar{R}$ $LCL = D_3\bar{R}$
$\bar{X}$ and s Average and Standard Deviation Chart	≥ 10	$\bar{\bar{X}} = \dfrac{sum\ of\ subgroup\ averages}{number\ of\ subgroups}$ $UCL = \bar{\bar{X}} + A_3\bar{s}$ $LCL = \bar{\bar{X}} - A_3\bar{s}$ $\bar{s} = \dfrac{sum\ of\ subgroup\ sigmas}{number\ of\ subgroups}$ $UCL = B_4\bar{s}$ $LCL = B_3\bar{s}$
X - $\overline{MR}$ Individuals and Moving Range Chart	1	$\bar{X}\ Centerline: \bar{X} = \dfrac{(X_1 + X_2 + ...X_k)}{k}$ $UCL = \bar{X} + E_2\overline{MR}$ $LCL = \bar{X} - E_2\overline{MR}$ $MR\ Centerline = \overline{MR}$ $UCL = D_4\overline{MR}$ $LCL = D_3\overline{MR}$

FIGURE 8.4
Equations for variable control charts.

TABLE 8.1

Control Chart Constants

Sample Observations	Chart for X-Bar		Chart for Std. Deviation				Chart for Ranges			
	Control Limit Factors		Centerline Factors	Control Limit Factors			Centerline Factors	Control Limit Factors		
n	A_2	A_3	C_4	B_3	B_4		E_2	D_3	D_4	
2	1.880	2.659	0.7979	0	3.267		2.660	0	3.267	
3	1.023	1.954	0.8862	0	2.568		1.772	0	2.574	
4	0.729	1.628	0.9213	0	2.266		1.457	0	2.282	
5	0.577	1.427	0.9400	0	2.089		1.290	0	2.114	
6	0.483	1.287	0.9515	0.030	1.970		1.184	0	2.004	
7	0.419	1.182	0.9594	0.118	1.882		1.109	0.076	1.924	
8	0.373	1.099	0.9650	0.185	1.815		1.054	0.136	1.864	
9	0.337	1.032	0.9693	0.239	1.761		1.010	0.184	1.816	
10	0.308	0.975	0.9727	0.284	1.716		0.975	0.223	1.777	

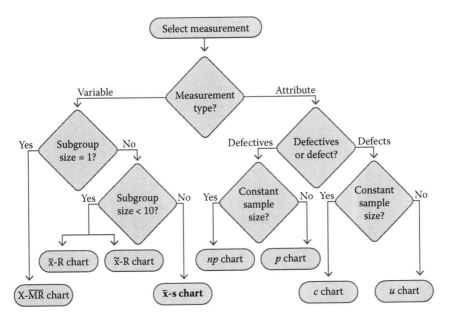

FIGURE 8.5
Location of X-bar and s chart.

The average and standard deviation chart (X-bar and s, respectively) is quite similar to the average and range chart, except that the statistic used to measure subgroup dispersion is the subgroup standard deviation instead of the subgroup range.

- Collect data by subgroup as shown in Figure 8.6.
- Calculate:
  - Mean for each subgroup
  - Standard deviation within each subgroup
  - The grand mean

See Figure 8.7.

Next Step:
- Calculate:
  - The mean of standard deviation values
  - Find the appropriate A3 value from the reference table
  - Calculate the UCL and LCL for the mean
  - Find the B4 and B3 values from the reference table
  - Calculate the UCL and LCL for standard deviation

Title:				Units:		
	Sample 1	Sample 2	Sample 3	Sample 4	Sample 5	
1	55	45	49	48	54	
2	48	51	53	45	48	
3	50	48	52	47	48	
4	45	53	55	45	56	
5	46	45	55	54	51	
Mean						
STD Dev						
Grand mean$_{\bar{\bar{x}}}$			STD Dev mean$_{\bar{s}}$			
UCL$_{\bar{x}}$			UCL$_{\bar{s}}$			
LCL$_{\bar{x}}$			LCL$_{\bar{s}}$			

FIGURE 8.6
X-bar s chart data sheet.

Title:				Units:		
	Sample 1	Sample 2	Sample 3	Sample 4	Sample 5	
1	55	45	49	48	54	
2	48	51	53	45	48	
3	50	48	52	47	48	
4	45	53	55	45	56	
5	46	45	55	54	51	
Mean	48.8	48.6	52.8	47.8	51.4	
STD Dev	4.0	3.4	2.5	3.7	3.6	
Grand mean		49.9		STD Dev mean		
UCL$_{\bar{x}}$				UCL$_{\bar{s}}$		
LCL$_{\bar{x}}$				LCL$_{\bar{s}}$		

FIGURE 8.7
X-bar s chart calculation of grand mean.

Title:				Units:		
	Sample 1	Sample 2	Sample 3	Sample 4	Sample 5	
1	55	45	49	48	54	
2	48	51	53	45	48	
3	50	48	52	47	48	
4	45	53	55	45	56	
5	46	45	55	54	51	
Mean	48.8	48.6	52.8	47.8	51.4	
STD Dev	4.0	3.4	2.5	3.7	3.6	
Grand mean		49.9		STD Dev mean		3.4
UCL$_{\bar{x}}$		54.8		UCL$_{\bar{s}}$		7.1
LCL$_{\bar{x}}$		45.0		LCL$_{\bar{s}}$		0

FIGURE 8.8
X-bar and s chart calculation of control limits.

See Figure 8.8 above.

To graph the X-bar and s, do the following:
- Use the grand mean to draw the center.
- Draw a line to display the UCL and LCL.
- Plot the subgroup means.
- Connect each point to form a line graph as shown in Figure 8.9.
- Repeat the procedure to calculate and graph the standard deviation range graph as done in Figure 8.10.

See Figure 8.11 for the completed X-bar and s graph.

Summary of X-Bar and s

The X-bar and s (average and standard deviation) chart is not used nearly as much as the X-bar and R chart.

Advantages are
- When the subgroup sizes are fairly large (greater than 10), it is often beneficial to consider the average and standard deviation

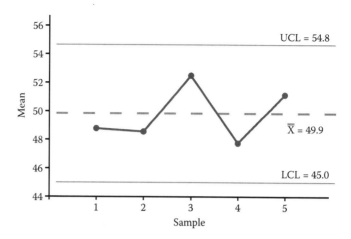

FIGURE 8.9

Draw control limits and grand mean centerline and plot sample means.

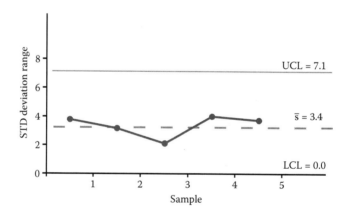

FIGURE 8.10

Draw control limits and S-bar centerline; plot sample S-values.

chart, since using the range as the measure of dispersion may not yield a good estimate of process variability.

- It may also be used when more sensitivity in detecting a process shift is desired, as in the case where the product being manufactured is quite expensive and any change in the process could either cause quality problems or add unnecessary costs.

Disadvantages are

- False signals may be issued at a much higher rate than for other types of control charts.
- The X-bar and s chart is complex to construct and use.

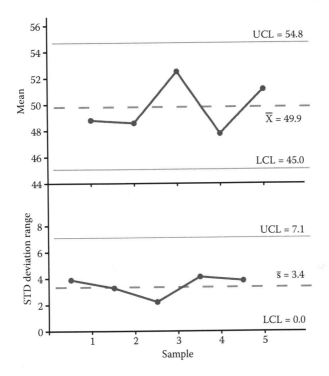

FIGURE 8.11
Completed X-bar and s graph.

X-MR or I-XR Chart

See the highlighted position in the road map in Figure 8.12. Use this chart with limited data, such as when production rates are slow, testing costs are very high, or there is a high level of uncertainty relative to future projects.

This chart is suitable for

- Individual measurements or situations where there is no basis for rational subgrouping
- Infrequently repeating processes
- Processes operating differently at different times

X-MR charts are applicable to situations when the sample size is $n = 1$. Examples include

- The early stages of a process when one is not quite sure of the structure of the process data
- Monthly data

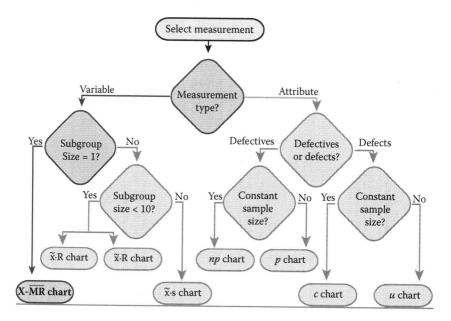

FIGURE 8.12
Location of X-MR and I-XR charts.

- When analyzing every unit (thus no basis for rational subgrouping)
- Slow production rates with long intervals between observations
- When differences in measurements are too small to create an objective difference
- When measurements differ only because of laboratory or analysis error
- Taking multiple measurements on the same unit (as thickness measurements on different places of a sheet of aluminum)

The procedure for creating this chart using the formulas in Figure 8.13 and data sheet in Figure 8.14 is as follows:

- Collect data.
- Calculate range (MR).
- Calculate mean for data.
- Find E_2 constant.
- Calculate UCL and LCL for mean and MR mean.

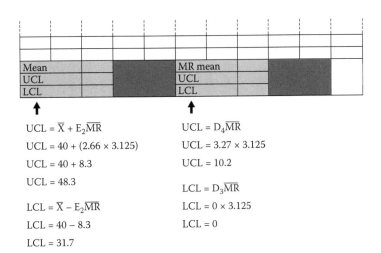

$$UCL = \overline{X} + E_2\overline{MR}$$

$$UCL = 40 + (2.66 \times 3.125)$$

$$UCL = 40 + 8.3$$

$$UCL = 48.3$$

$$LCL = \overline{X} - E_2\overline{MR}$$

$$LCL = 40 - 8.3$$

$$LCL = 31.7$$

$$UCL = D_4\overline{MR}$$

$$UCL = 3.27 \times 3.125$$

$$UCL = 10.2$$

$$LCL = D_3\overline{MR}$$

$$LCL = 0 \times 3.125$$

$$LCL = 0$$

FIGURE 8.13
Formulas for X-MR and I-XR charts.

Title:							Units:		
	Aug 1	Aug 2	Aug 3	Aug 4	Aug 5	Aug 6	Aug 7	Aug 8	Aug 9
Data	39	42	37	39	44	40	38	41	40
MR		3	5	4	5	4	2	3	1
Mean	40				MR mean		3		
UCL	48				UCL		9.8		
LCL	32				LCL		0		

FIGURE 8.14
Data sheet for X-MR or I-XR chart.

Draw control limits and mean centerline and plot sample means as shown in Figure 8.15.

Advantages of the X-MR chart are

- It is useful even in situations with small amounts of data.
- It is easy to construct and apply.
- It is useful in the early stages of a new process when not much is known about the structure of the data.

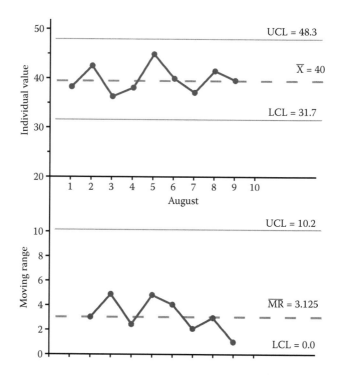

FIGURE 8.15
Draw control limits and mean centerline and plot sample means.

The disadvantages is that the X-MR chart cannot discern between common-cause and special-cause variations.

Attribute Data Control Charts

The formulas for attribute data charts are shown in Figure 8.16.

When working with attribute data charts, we need to understand the difference between two related terms:

Defect: An undesirable result on a product; also known as nonconformity.
Defective: An entire unit failing to meet specifications; also known as a nonconformance.

Note: A unit may have multiple defects.

Attribute Data Charts

CHART TYPE	SAMPLE SIZE	CONTROL LIMITS
p Chart Fraction Defective	Variable	$Centerline : \bar{p} = \dfrac{\sum np}{\sum n}$ $UCL = \bar{p} + 3\sqrt{\dfrac{\bar{p}(1 - \bar{p})}{n}}$ $LCL = \bar{p} - 3\sqrt{\dfrac{\bar{p}(1 - \bar{p})}{n}}$
np Chart Number Defective	Constant	$Centerline : n\bar{p} = \dfrac{\sum np}{k}$ $UCL = n\bar{p} + 3\sqrt{n\bar{p}(1 - \bar{p})}$ $LCL = n\bar{p} - 3\sqrt{n\bar{p}(1 - \bar{p})}$
c Chart Number of Defects	Constant	$Centerline : \bar{c} = \dfrac{\sum c}{k}$ $UCL = \bar{c} + 3\sqrt{\bar{c}}$ $LCL = \bar{c} - 3\sqrt{\bar{c}}$
u Chart Number of Defects Per Unit	Variable	$Centerline : \bar{u} = \dfrac{\sum c}{\sum n}$ $UCL = \bar{u} + 3\sqrt{\dfrac{\bar{u}}{n}}$ $LCL = \bar{u} - 3\sqrt{\dfrac{\bar{u}}{n}}$

FIGURE 8.16
Formulas for attribute data charts.

p Charts

The *p* chart is one of the most used types of attribute charts. It shows the proportion of defective items in successive samples of equal or varying size. The most often used *p* chart employs fraction nonconforming data. It provides an estimate of the ongoing quality level, and it is easy to use. A customer might request using a *p* chart to ensure that a certain quality level is being obtained. It may be noted that p charts can be used when the subgroup size varies.

See the highlighted position of *p* chart in the road map in Figure 8.17.

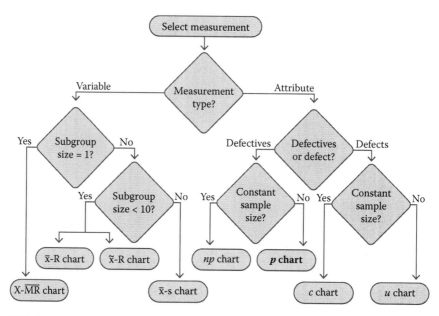

FIGURE 8.17
Location of p chart.

To make a p chart for an example with varying sample sizes, do the following.

- Refer to the p chart formula.
- Calculate:
  - The totals
  - p value (proportion) of each shipment
  - p-bar for central line
  - UCL and LCL

Collect the data as shown in Figure 8.18.
The finished data sheet table is shown in Figure 8.19.

p-bar = $\Sigma p/12$, where p is the fraction of bruised apples in one crate (e.g., 37/200 = 0.185). Note: The graph shows the rounded figure of 0.190.
n-bar = $\Sigma n/12$, where n is the number apples in one crate (e.g., 2800/12 = 233).

Draw the p chart as shown in Figure 8.20.

Shipping #	Crates Shipped	Apples Shipped	Bruised Apples	P
1	1	200	37	
2	1	200	33	
3	1	200	44	
4	1	200	41	
5	2	400	75	
6	2	400	81	
7	1	200	35	
8	1	200	47	
9	1	200	30	
10	1	200	36	
11	1	200	33	
12	1	200	40	
Total				

Values	
$\bar{n}$	
$\bar{p}$	
UCL	
LCL	

FIGURE 8.18
Data sheet for p chart.

Shipping #	Crates Shipped	Apples Shipped	Bruised Apples	P
1	1	200	37	0.185
2	1	200	33	0.165
3	1	200	44	0.220
4	1	200	41	0.205
5	2	400	75	0.188
6	2	400	81	0.203
7	1	200	35	0.175
8	1	200	47	0.235
9	1	200	30	0.150
10	1	200	36	0.180
11	1	200	33	0.165
12	1	200	40	0.200
Total	14	2800	532	0.189

Values	
n-bar	233
p-bar	0.189
UCL	0.266
LCL	0.112

FIGURE 8.19
Calculation of p-bar, n-bar, and control limits.

np Chart

See the highlighted position of *np* chart in the road map in Figure 8.21.

The example of a data record sheet for *np* chart is shown in Figure 8.22. The *np* chart records the number of defective units (nonconformances) and is more difficult to use when the subgroup size varies.

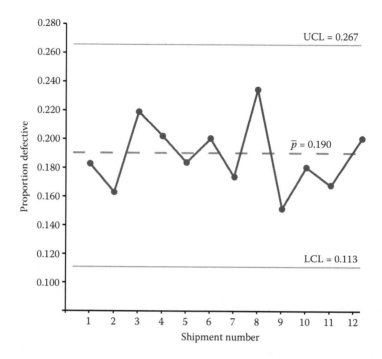

FIGURE 8.20
Drawing of a *p* chart.

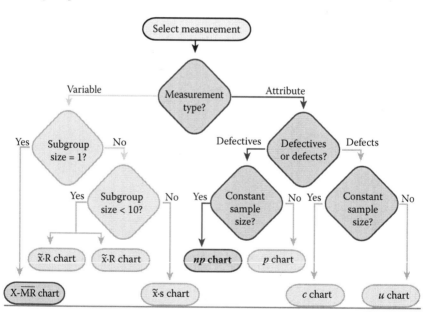

FIGURE 8.21
Location of *np* chart.

Lot	Defects	Sample Size
1	37	1000
2	33	1000
3	44	1000
4	41	1000
5	35	1000
6	47	1000
7	30	1000
8	36	1000
9	33	1000
10	40	1000
Total	**376**	**10,000**

Values	
$n\bar{p}$	37.6
$\bar{p}$	0.0376
UCL	55.6
LCL	19.6

FIGURE 8.22
Data sheet to calculate np-bar and p-bar.

MATH SOLUTION

$$\bar{np} = \frac{\Sigma\ nonconformances}{\#\ subgroups}$$

$$\bar{np} = \frac{376}{10}$$

$$\bar{np} = 37.6$$

$$\bar{p} = \frac{\Sigma\ noncomformances}{\Sigma\ subgroup\ units}$$

$$\bar{p} = \frac{376}{10,000}$$

$$\bar{p} = .0376$$

$$UCL = n\bar{p} + 3\sqrt{n\bar{p}(1-\bar{p})}$$

$$UCL = 37.6 + 3\sqrt{37.6(1-0.0376)}$$

$$UCL = 37.6 + 3\sqrt{36.18624}$$

$$UCL = 37.6 + (3 \times 6.0155)$$

$$UCL = 37.6 + 18.0$$

$$UCL = 55.6$$

$$LCL = n\bar{p} - 3\sqrt{n\bar{p}(1-\bar{p})}$$

$$LCL = 37.6 - 3\sqrt{37.6(1-0.0376)}$$

$$LCL = 37.6 - 18.0$$

$$LCL = 19.6$$

FIGURE 8.23
Calculation of UCL and LCL for np chart.

- Refer to the np chart formula (Figure 8.23).
- Calculate:
  - The totals.
  - p value (proportion) of each shipment.
  - p-bar for the central line.
  - UCL and LCL.

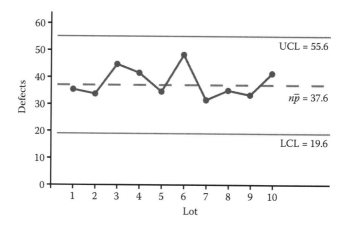

FIGURE 8.24
Completed *np* chart.

In Figure 8.22, use the data sheet to calculate *np*-bar and *p*-bar:

np-bar = 376/10 lots = 37.6
p-bar = 376/10,000 = 0.376

Draw the *np* chart as shown in Figure 8.24.

c and u Charts

These attribute control charts are used to monitor the count of individual defects rather than the number of defective units. They are used where there are opportunities for many defects per defined inspection unit.

With *c* and *u* charts, it is very important to define the defects and the unit.

The unit is the area of opportunity to count the defects.

Inspection Unit	Type of Defects Counted
50 miles of pipeline	Weld defects
10 yards of cloth	Blemishes, snags
50 circuit boards	Solder joint defects, damaged components
100 forms	Incorrect data entry, missing data

You can create *c* and *u* charts to plot attribute data where you are tracking the number of defects per unit. Use c charts to plot the total quantity of defects based on samples of equal size, for example, the number of recordable injuries in a plant each month. As another example, if the inspection unit is 100 forms, count the defects on a sample of 100 forms and plot that number on the *c* chart.

Use *u* charts to plot the defects per unit based on samples of equal or unequal size.

c Chart

The *c* chart formula assumes counting the number of defects in the same area of opportunity. The *c* in the formula is the number of defects found in the defined inspection unit, and that is plotted on the chart.

The *c* chart monitors the number of nonconformities (defects) and requires the inspection unit to be defined clearly and the area of opportunity to be consistent.

The *c* chart formula:

$$\bar{c} = \frac{\sum c}{k}$$

See the highlighted position of *c* chart in the road map in Figure 8.25.

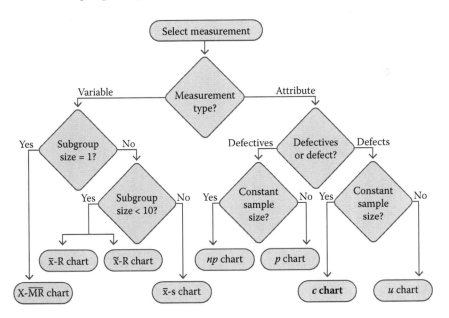

FIGURE 8.25
Location of *c* chart.

An example of data sheet for *c* chart is shown in Figure 8.26.

- Collect the data.
- Calculate all values as per the formulas for *c* chart and the data sheet as shown in Figure 8.27.
- Draw the chart *c* chart as shown in Figure 8.28.

Lot	Defects
1	8
2	7
3	14
4	10
5	3
6	9
7	2
8	12
9	11
10	10
Total	

Values	
$\bar{c}$	
UCL	
LCL	

FIGURE 8.26
Data sheet for *c* chart.

c chart   Constant
Number of Defects

$$Centerline: \bar{c} = \frac{\sum c}{k}$$

$$UCL = \bar{c} + 3\sqrt{\bar{c}}$$

$$LCL = \bar{c} - 3\sqrt{\bar{c}}$$

Lot	Defects
1	8
2	7
3	14
4	10
5	3
6	9
7	2
8	12
9	11
10	10
Total	86

Values	
$\bar{c}$	8.6
UCL	17.4
LCL	0

FIGURE 8.27
Calculation of UCL and LCL for *c* chart.

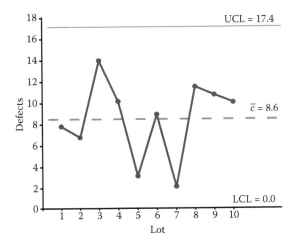

FIGURE 8.28
Completed *c* chart.

u Chart

With a *u* chart, the number of inspection units may vary. The *u* chart requires an additional calculation with each sample to determine the average number of defects per inspection unit. The *n* in the formula is the number of inspection units in the sample.

The *u* chart monitors the defects (nonconformities) per unit when the number of inspection units is allowed to vary. It essentially changes the counts into rates in cases where the area of opportunity varies from sample to sample.

See the highlighted position of *u* chart in the road map in Figure 8.29.

- Collect data by sample using the form and formulas in Figure 8.30.
  - Total rolls shipped: 14
  - Total defects: 532
- Calculate the *u*-bar for all data:

$$\bar{u} = \frac{\sum c}{\sum n}$$

$$\bar{u} = 532/14 = 38$$

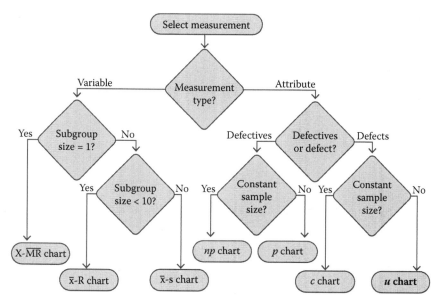

FIGURE 8.29
Location of *u* chart.

Shipping #	Rolls	Inspected Defects	Defects per Roll
1	1	37	37
2	1	33	33
3	1	44	44
4	1	41	41
5	2	75	37.5
6	2	81	40.5
7	1	35	35
8	1	47	47
9	1	30	30
10	1	36	36
11	1	33	33
12	1	40	40
Total	14	532	

$\bar{u}$	
UCL n = 1	
LCL n = 1	
UCL n = 2	
LCL n = 2	

u chart   Variable
Number of defects
Per unit

$$Centerline: u = \frac{\sum c}{\sum n}$$

$$UCL = \bar{u} + 3\sqrt{\frac{\bar{u}}{n}}$$

$$LCL = \bar{u} - 3\sqrt{\frac{\bar{u}}{n}}$$

FIGURE 8.30
Data sheet and formula for *u* chart.

- Calculate UCL and LCL for $n = 1$ and $n = 2$ using the formulas given in Figure 8.16.
- Example: UCL and LCL for n = 1 and n = 2
  - UCL and LCL data for u chart is shown in Figure 8.31.
  - Plot the chart using the data of Figure 8.31.
  - The completed charts are shown in Figures 8.32 and 8.33.

Figure 8.34 shows the calculations in Math for u chart.

Shipping #	Rolls	Inspected Defects	Defects per Roll
1	1	37	37
2	1	33	33
3	1	44	44
4	1	41	41
5	2	75	37.5
6	2	81	40.5
7	1	35	35
8	1	47	47
9	1	30	30
10	1	36	36
11	1	33	33
12	1	40	40
Total	14	532	

$\bar{u}$	38.0
UCL n = 1	56.5
LCL n = 1	19.5
UCL n = 2	51.1
LCL n = 2	24.9

FIGURE 8.31
UCL and LCL data for u chart.

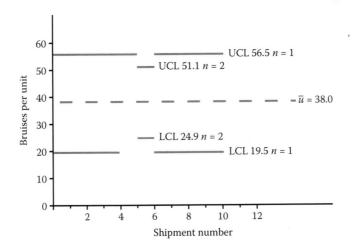

FIGURE 8.32
Plot for u-bar centerline and UCL and LCL for $n = 1$ and $n = 2$, respectively.

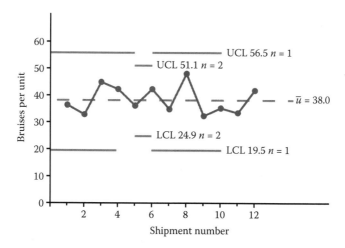

FIGURE 8.33
Completed *u* chart.

Summary of c and u Charts

Use the *c* chart to monitor the number of defects (nonconformities). The application of the *c* chart requires the inspection unit to be defined clearly and the areas of opportunity to be consistent. The *u* chart monitors the defects (nonconformities) per unit and does not require a constant sample size. This is the reason you had to calculate two sets of limits in the example. It essentially changes the counts into rates in cases where the area of opportunity varies from sample to sample. The Poisson model is the statistical model that is the foundation of *c* charts and *u* charts. Like the binomial model for *p* charts and *np* charts, the Poisson model has several conditions that must be met:

- The counts must be discrete events.
- The counts must be clearly defined with an unambiguous area of opportunity described.
- The events must be independent.
- The defects (nonconformities) must be few compared to the areas of opportunity.

The advantage is that they can be used where the nonconformities from many potential sources may be found in a single inspection.

The disadvantage is that they require a constant sample size.

Shipping #	YDS MATL INSPECTED	DEFECTS	DEFECTS per UNIT
1	1	37	37
2	1	33	33
3	1	44	44
4	1	41	41
5	2	75	37.5
6	2	81	40.5
7	1	35	35
8	1	47	47
9	1	30	30
10	1	36	36
11	1	33	33
12	1	40	40
Total	14	532	

$\bar{u}$	38.0
UCL n=1	56.5
LCL n=1	19.5
UCL n=2	51.1
LCL n=2	24.9

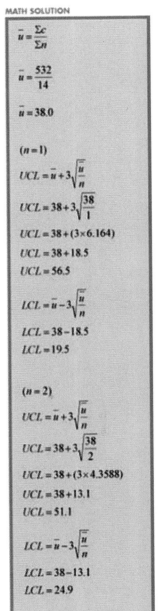

MATH SOLUTION

$$\bar{u} = \frac{\Sigma c}{\Sigma n}$$

$$\bar{u} = \frac{532}{14}$$

$$\bar{u} = 38.0$$

$(n=1)$

$$UCL = \bar{u} + 3\sqrt{\frac{u}{n}}$$

$$UCL = 38 + 3\sqrt{\frac{38}{1}}$$

$$UCL = 38 + (3 \times 6.164)$$

$$UCL = 38 + 18.5$$

$$UCL = 56.5$$

$$LCL = \bar{u} - 3\sqrt{\frac{u}{n}}$$

$$LCL = 38 - 18.5$$

$$LCL = 19.5$$

$(n=2)$

$$UCL = \bar{u} + 3\sqrt{\frac{u}{n}}$$

$$UCL = 38 + 3\sqrt{\frac{38}{2}}$$

$$UCL = 38 + (3 \times 4.3588)$$

$$UCL = 38 + 13.1$$

$$UCL = 51.1$$

$$LCL = \bar{u} - 3\sqrt{\frac{u}{n}}$$

$$LCL = 38 - 13.1$$

$$LCL = 24.9$$

FIGURE 8.34
Math for *u* chart.

CONTROL PLAN

Introduction

The control plan is the centralized document to keep track of the status of all significant process characteristics.

A control plan form along with a description of each of the columns (which can be modified to fit your needs) is shown in Figure 8.35.

Critical-to-quality characteristic (CTQC): End product characteristic proven to be important to the customer, along with hierarchical reference number.

Significant characteristic number: Reference number to organize significant characteristics within a hierarchy that relates to the corresponding CTQCs.

Significant characteristic description: Process characteristics that have a significant impact on the CTQC.

Chart type: X-bar and R chart, *p* chart, *c* chart, and trend chart.

Chart champion: Name of the process owner.

Chart location: Location where the cart is kept.

Measurement method: Method used to collect the measurement data, for example, scale or caliper.

Measurement study: Yes or no to denote whether a measurement system analysis has been completed. If yes, show percent total error.

Reaction plan: Reference number to a reaction plan flowchart that tells the data plotter what to do in the event of an out-of-control or out-of-specification condition. Reaction plans may be somewhat generic for families of process with similar criticality.

Gage number: Reference number for the gage that corresponds to the calibration tracking system.

Sampling plan: How many samples are drawn at what frequency?

Process stability: Is the process in a state of statistical control—yes or no.

Cp/Cpk: If process is stable, calculate Cp and Cpk.

Control plan												Revision number		1.0
		Organization:										Date		
		Location:												
Critical-to-Quality Characteristic	Sig. Char. #	Significant Characteristic Description	Chart Type	Chart Champion	Chart Location	Measurement Method	Measurement Study	Reaction Plan	Gage Number	Sampling Plan	Process Stability	Cp/Cpk			
A	A-1													
	A-2													
B	B-1													
	B-2													

FIGURE 8.35
Control plan form.

SUSTENANCE OF IMPROVEMENTS

Lessons Learned

A completed Six Sigma project generates a wealth of information. Lessons learned is a process to capture, document, and share lessons learned, thus infusing change in the organization.

The process answers the following:

- What went well?
- What could have been done differently?
- What could be improved?
- What did we do that we should not have?
- Did all our various stakeholders interact efficiently and effectively?
- Where were there gaps?
- Where were there overlaps?
- What can be done differently next time to make the situation easier for all parties involved?

The documented process promotes continued improvement and identification of additional opportunities by

- Enabling others to learn how the project was planned, implemented, and monitored
- Helping resolve issues
- Allowing resources to be tracked back to their work in the project
- Creating an audit trail
- Providing direction to revise or revive the project later

Implementation of Training Plan

It is important to note that the training of the personnel responsible for executing the process is a critical element for sustaining the gains of improved processes.

Training is often overlooked or viewed as an unnecessary cost or imposition on the time of already overworked employees.

There are two types of training:

1. Initial
2. Recurring

Initial training is used for existing employees who have been responsible for executing the old process and will continue to be responsible for executing the new process.

It also addresses the needs of employees who are new to the process. It may be conducted off the job, on the job, or both. The correct combination depends on the skill levels required, process complexity, and experience levels, among other factors.

Initial training serves to calibrate employees and minimize variation in how the process is performed.

Recurring training is used to minimize deterioration in the process performance over time. Deterioration usually occurs as employees become comfortable with the new process and fail to recognize minor changes or tweaks to the process.

This happens through

- Carelessness
- Poor documentation
- Perhaps even inadequate process document control management that permits obsolete documentation to remain in the process stream

In addition, on-the-job training is another factor that often adversely impacts process performance.

Specific facts, details, and nuances are frequently left out as they pass from worker to worker.

Recurring training restores the process execution to its original design.

The frequency of recurring training should be determined based on process metrics and employee performance. Such training may be offered at specified intervals or conducted as required to serve the needs of underperforming employees.

Standard Operating Procedures and Work Instructions

In Book 2 we have discussed the importance of standard work. Documented standard work in Lean thinking is the same as documented standard operating procedures (SOPs) and work instructions. The purpose of these documents is to make certain that the activity is performed in the same way over time. This is especially important when multiskilled, cross-trained personnel move into a variety of positions. The other main purpose of these documents is to help reduce process variation. The development and updating of these documents must involve the people who perform the work. The documentation process is begun by listing major steps. Succeeding iterations examine the steps from the previous one and break them into smaller substeps. This process can continue until further deconstruction is not useful.

Two main considerations regarding documents are

- Documents must be kept current. In an environment of continuous improvement, processes may be continuously changing. As a result, it is very possible that different documentation releases for the same documentation may be in use simultaneously. Therefore, it is vital that employees always have access to the appropriate documentation based on the effectiveness of the change. It is easy to see how the complexity of managing such a documentation system can grow exponentially. With today's web-based and other technologies, documentation configuration management systems can be designed and developed to ensure that the proper documentation is available for the right process on the right part at the right time.

- Multiple formats for documentation exist. The right choice depends on how the documentation is to be used, by whom, and at what skill level. However, as we have seen in the visual factory, documentation best practices suggest developing documents that are color coded, light on words, and rely heavily on graphics, illustrations, and photographs. Photographs of acceptable and unacceptable products and procedures have been found to be very useful. Some organizations have had success with the use of videotapes and DVDs for depicting complex operations. As always, the level of detail provided in any set of documentation should be reflective of the skills and education levels of the personnel doing the actual work and the degree to which variation must be controlled.

Remember, robust processes usually require less detailed documentation than nonrobust processes.

Ongoing Evaluation

Once a process has been improved, it must be monitored to ensure the gains are maintained and to determine when additional improvements are required.

The following tools and methods are available to assist ongoing evaluation:

- **Control charts:** These are used to monitor the stability of the process and to determine when special cause is present and when to take appropriate action. The choice of a particular control chart depends on the nature of the process. When out-of-control conditions occur, action is required to restore stability. Control charts represent the voice of the process.
- **Process capability studies:** These studies provide us with the opportunity to understand how the voice of the process (i.e., control limits) compares with the voice of the customer (i.e., specifications), and help us determine whether the process average must be shifted or recentered or the variation reduced.
- **Process metrics:** This includes a wide variety of in-process and end-of-process metrics that measure its overall efficiency and effectiveness.

Examples include cycle times, takt time, take rate, work in process, backlogs, defect rates, rework rates, and scrap rates.

Taken collectively, the above tools and methods help us gage the overall health of a process, and provide triggers for reevaluating a process for further improvement.

Note: Take rate refers to the percentage of visitors who took interest in an action but did not actually follow through with that action.

So, if you have 3000 web site visitors and 600 go to the download page, your take rate is 20%. If 200 download the article your conversion rate is 6.6%.

9

Design for Six Sigma

DFSS AND SIX SIGMA

Design for Six Sigma (DFSS) and Six Sigma have overlapping functions as shown in Figure 9.1. DFSS originated in the U.S. Department of Defense and the National Aeronautics and Space Administration (NASA). Like Six Sigma, it seeks to understand the customer's perceptions, needs, and expectations that are critical to ensure a successful product.

Additionally, DFSS is a proactive, rigorous, and systematic method using tools, training, and measurements for integrating customer requirements into the product development process.

By preventing defects, DFSS increases the probability of success (reliability) and the robustness of the product much earlier by applying scientific methods like

- Quality functional deployment (QFD)
- Failure mode and effects analysis (FMEA)
- Robust design optimization (RDO) using
    - Design of experiments (DOE) and other techniques to identify the most robust design solutions for customer usage. Using RDO processes, manufacturers can identify errors early in the design process, when they are easy and less costly to fix.
    - Parameter diagrams, such as RDO and P diagrams, to incorporate robustness and reliability into the product design while keeping a control on costs.

As shown in Figure 9.1, DFSS methodologies integrate tools, methods, processes, and team members from the start and throughout the product and process design.

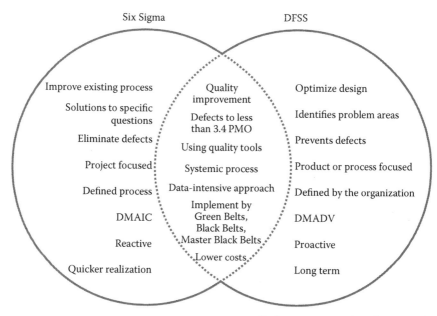

Six Sigma   DFSS

Improve existing process   Quality improvement   Optimize design

Solutions to specific questions   Defects to less than 3.4 PMO   Identifies problem areas

Eliminate defects   Using quality tools   Prevents defects

Project focused   Systemic process   Product or process focused

Defined process   Data-intensive approach   Defined by the organization

DMAIC   Implement by Green Belts, Black Belts, Master Black Belts   DMADV

Reactive   Proactive

Quicker realization   Lower costs   Long term

Define, measure, analyze, design, verify

FIGURE 9.1
Venn diagram showing Six Sigma and DFSS functions.

The following tools and processes are involved in DFSS methodologies:

Tools:

- DMAIC tools: Basic statistics, analysis of variance (ANOVA), DOE, measurement system analysis, process capability, process mapping, and hypothesis testing
- Design tools: Critical to quality (CTQ), QFD, robust design, tolerance analysis, TRIZ, and scorecards

Business processes:

- Project management, cost analysis, strategic planning, risk management, and benchmarking
- Technical processes: Risk analysis, design reviews, prototyping, pilot runs and prototype production, and simulation

DFSS AND ROI

A well-developed DFSS project development plan will help to ensure success and maximize the return on investment (ROI).

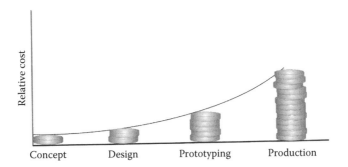

FIGURE 9.2
Product life cycle cost.

It is believed that 70% to 80% of quality problems are regarded as design related. This means that the design engineers and other process designers have the best opportunity to improve product quality and save costs.

As a process moves through the phases of concept to design, prototyping, and production, the cost of resource inputs increases, and so does the cost of eliminating quality problems. The greater the initial investment in DFSS methodology to eliminate design issues, the lower are the life cycle costs associated with the process (Figure 9.2).

Business challenges addressed by DFSS include

- Customers demanding product or process excellence
- Managing variability
- Improving predictability and capability
- Reducing costs and increasing profitability
- Increasing development effectiveness
- Managing development costs
- Meeting customer requirements

DFSS METHODOLOGIES

The main topics and tools are

- Define, measure, analyze, design, and verify (DMADV)
- Define, measure, analyze, design, optimize, and verify (DMADOV)
- DFSS teams
- Design for X (DFX)

- Reliability
- Tolerance

Special design tools include

- Porter's five forces analysis
- TRIZ

DMADV

As DMAIC guides the Six Sigma Black Belt process, the process stages for DFSS are define, measure, analyze, design, and verify (DMADV) (Figure 9.3). It is used when designing a new process, product, service, or transaction. DMADV works closely with obtaining information and analyzing the voice of the customer in order to meet customer requirements. DFSS contains such a large body of knowledge for each stage that DFSS is a course in itself. The purpose of this section is to provide an overview of DFSS.

DMADOV

The DMADOV process steps are outlined as follows:

Define: Define the project (define goals, identify customers, and identify customer expectations).
Measure: Measure the opportunity (determine customer needs and specifications and benchmark).

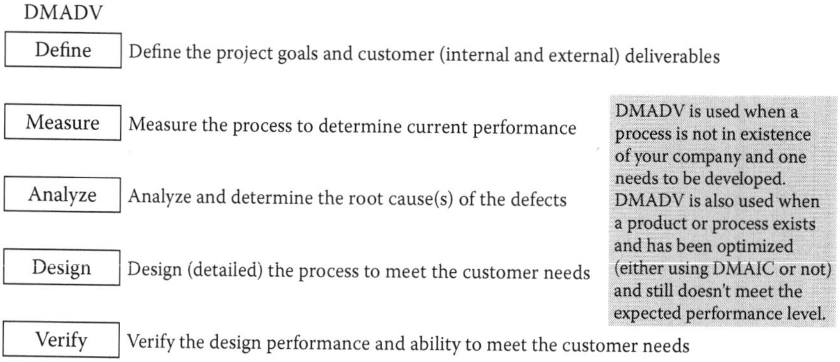

DMADV

Define	Define the project goals and customer (internal and external) deliverables
Measure	Measure the process to determine current performance
Analyze	Analyze and determine the root cause(s) of the defects
Design	Design (detailed) the process to meet the customer needs
Verify	Verify the design performance and ability to meet the customer needs

DMADV is used when a process is not in existence of your company and one needs to be developed. DMADV is also used when a product or process exists and has been optimized (either using DMAIC or not) and still doesn't meet the expected performance level.

FIGURE 9.3
DMADV process.

Analyze: Analyze the process options to meet customer need.

Design: Design the process by developing a detailed process and experiments to prove the design meets customer needs.

Optimize: Optimize the process by testing the new process and redesigning as necessary to meet customer specifications.

Validate: Validate the performance and ability of the process to meet customer needs and deploy the new process.

DFSS Teams

When a team develops a new design from start to finish, engineering should not be the only department involved. DFSS teams can derive important information using cross-functional team members from other departments, including

Finance
Marketing
Sales
Human resources
Engineering
Supply chain management
Information technology
Legal
Environmental protection

The following inputs benefit the DFSS projects and are critical to the smooth operation and success of DFSS projects:

- Learning and training needs from human resources
- Patent infringement advice from the legal department
- Money matters from the finance department
- Customer requirements from marketing and sales
- Sound supply chain management from purchasing
- Product design from
    - Subject matter experts (SMEs)
    - Six Sigma Black Belts and Green Belts as team leaders
    - Master Black Belts as mentors and trainers
    - Project champions as process owners and roadblock breakers

There are a wide variety of examples showing when to use DFSS and DMAIC:

1. An investment company wants to attract more young adult investors: DFSS.
2. A company uses customer feedback to initiate a software update: DMAIC.
3. Problems and issues in current processes are addressed: DMAIC.
4. A hospital desires to implement a new treatment procedure: DFSS.
5. A firm decides redesign is cheaper than problem solving: DFSS.

Design for X

DFX is an approach for designing products and services that meet customer requirements.

It is a cross-functional team design activity involving manufacturing, distribution, and service organizations.

DFX strategy reviews design continually to find ways to improve the product. For instance, when considering serviceability and maintenance, service personnel are involved who address their requirements and concerns.

Due to its use of cross-functional teams and the nature of continual review, DFX is needed within concurrent engineering (simultaneous engineering) as an approach to improve new product development where the product and associated processes are developed in parallel.

Concepts and Techniques for DFX

Design for Cost

- Also called value engineering
- Must consider price limitations

Design for Manufacturability

- Has the goal to design products and processes in such a way that they result in fewer problems during manufacturing
- Emphasizes robustness rather than ideal performance
- Reduces the probability of mistakes by reducing complexity
- Designs preventive mechanisms for likely errors, such as mistake proofing or Poka Yoke
- Reduces the number of parts
- Reduces the number of manufacturing operations

Design for Assembly

- Simplifies the product into fewer parts
- Makes the product easy to assemble
- Reduces service, decreases time to market, and reduces repair time

Sometimes design for manufacturability and design for assembly are combined (DFM/A).

The benefits of integrating DFM and DFA are

- Simpler designs
- Fewer parts
- Reduced assembly time
- Reduced production cost
- Fewer errors
- Fewer suppliers
- Easier to test and maintain

Design for Producibility (DFP)

- Is a key metric of the success of a product design when influencing design and concurrent engineering
- Identifies the needs of innovative manufacturing
- Ensures proposed process will satisfy design requirement
- Decreases cost
- Reduces concept-to-build cycle times
- Reduces risk

Design for Test

- Is important during development, production, and use
- Makes access points easily assessable
- Creates built-in test points
- Uses standard connections and interfaces
- Tests with standard equipment
- Develops a build-in self-test
- Prevents test escapes

Design for Serviceability and Safety

Serviceability

- Is important to the customer
- Designs easy service

- Makes access points simple, yet secure
- Assures reliability of individual components
- Balances reliability and cost with the product's intended use and life
- Reduces downtime for maintenance
- Reduces the number of maintenance tasks
- When applicable, uses disposable parts instead of parts requiring repair
- Eliminates or reduces the need for adjustment
- Uses mistake-proof fasteners and connectors

Safety

- Eliminates potential failure elements that may occur during operation
- Emphasizes safety throughout the product life: safe to manufacture, safe to sell, safe to use, and safe to dispose

Others

- Design for user-friendliness
- Design for ergonomics
- Design for appearance
- Design for packaging
- Design for features
- Design for time to market
- Design for environment

Concepts for DFX Diagram

See Figure 9.4.

Reliability

Failure is a real and important customer concern, whether it is software, sensitive medical equipment, a complex machine for an industrial process, or a kitchen appliance. The concerns are

- How long will the product work?
- How many failures will occur within so many years?
- How do different environmental conditions affect performance?

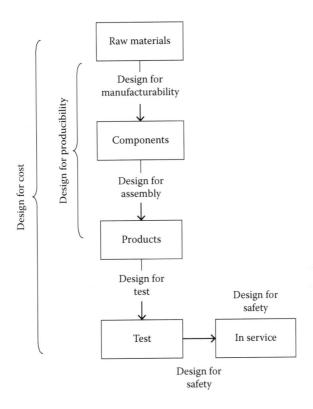

FIGURE 9.4
DFX concept diagram.

Since successful performance is important to customers, producers must determine how reliable the product or service will be.

Reliability is defined as the probability that a product will perform as stated, under specified operating conditions, for a given time period.

This definition contains four concepts:

1. Probability
   - This is the chance that a given outcome will occur.
   - It is a calculated, numerical value.
   - Probability theory provides the mathematical foundation.
2. Successful performance
   - This is a specifically defined set of criteria for goodness or failure.
   - Each unit's condition must be clearly defined; failure could mean total inoperativeness or diminished performance.
   - To calculate reliability, a product (or unit) exists in one of two states: successful performance or failure.

3. Operating conditions
- Operating conditions specify the environmental and use limits for operating the unit. For example, "this medicine must be stored in a dry room between 56 and 87°F."
- Customers have a responsibility to use a unit within these limits, but this is by no means a guarantee.
- Product designers must anticipate and design for stress conditions above proper use conditions.
4. Time
- Within the context of reliability, the time period involved must be specified.
- Times could be hours, years, miles, cycles, or some other measure tied to duration or amount of use.

Bathtub Curve

The model used to describe failure patterns of a product population over the entire life of the product is called the bathtub curve, named for its shape (Figure 9.5). The bathtub curve is a "plot" of failure rate versus time. It shows how fast product failures are taking place, and if the failure rate is increasing, decreasing, or staying the same. The curve is divided into three regions:

1. Region where burn-in failures and infant mortality failures take place
2. Region of stable operating life
3. Region of accelerated failure and wear-out failure

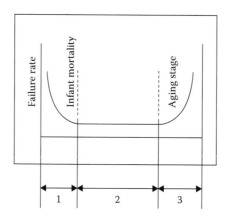

FIGURE 9.5
Bathtub curve.

Region Where Burn-In Failures and Infant Mortality Failures Take Place

This region is furthest to the left of the bathtub curve and is known as early life period. Failures during this period are referred to as infant mortality or early life failures. Systems in this phase of their life cycle are considered unsuitable for routine product operation and are not suitable for delivery to customers. The failures are caused by nonconformities introduced into a product by the production process.

Common sources of infant mortality include

- Inadequate materials
- Improper use
- Handling damage
- Overstressed components
- Improper setup or installation
- Power surges

Burn-in is the early life period and is sometimes called the burn-in period. Burn-in refers to the practice of running the system under conditions that simulate an operating environment for a period of time sufficient to allow the failure rate to stabilize. For many types of products, burn-in is performed at normal operating conditions. For others, burn-in is performed at higher-than-normal stress levels, such as increased temperatures and vibration levels.

During burn-in, many of the units containing nonconformities fail and are removed from the population. This improves the reliability of the units delivered to the customer and reduces the likelihood that a unit will fail in customer hands because of an early life cause. Unfortunately, the burn-in approach to ensuring high reliability is also very costly and can be less than 100% effective.

Even though virtually all new products experience early life failures, the failures are generally not used to make reliability predictions. Consequently, this section does not provide a method for calculating reliability during early life. However, it does address the relevance of quality engineering during this period.

Region of Stable Operating Life

This is the useful life period, and it has an exponential distribution. This region is the center of the bathtub curve and has several distinct characteristics:

- Where product use is intended
- Where customer use is the highest

- Where reliability engineers make the most reliability calculations and predictions
- Where a constant failure rate is reflected

The age of the product does not affect the probability of failure.

The failure rate in this region is extremely low if product reliability is to be high.

The exponential distribution can be used to model a product's times to failure.

Reliability for this region is quantified as mean time between failures (MTBF) for repairable product and mean time to failure (MTTF) for non-repairable product.

The formulas are as follows:

$$\text{MTBF} = T/R = [\theta] \tag{9.1}$$

where T is the total time and R is the number of failures,

$$\text{MTTF} = T/N \tag{9.2}$$

where T is the total time and N is the number of units under test, and

$$R(t) = e^{-\lambda t} \tag{9.3}$$

where $R(t)$ is the reliability estimate for a period of time, cycles, miles, and so forth (t); e is the base of the natural logarithms (2.718281828); and λ is the failure rate (1/MTBF or 1/MTTF).

Suppose 10 devices are tested for 500 hours. During the test, two failures occur.

The estimate of the MTBF is

$$T = \text{Total time} = 500 * 10 = 5000 \text{ hours}$$

$$R = 2$$

$$T/R = \text{MTBF} = 5000/2 = 2500$$

whereas MTTF = T/N = 5000/10 = 500.

If the MTBF is known, one can calculate the failure rate as the inverse of the MTBF.

The formula for the failure rate (λ) is

$$\lambda = \text{inverse of } \theta = 1/\theta = R/T$$

The formula for calculating the reliability is

$$R(t) = e^{-\lambda t}$$

A system in the useful life region has an MTTF of 5000 hours.

1. What is the failure rate of the system?
2. What is the reliability of the system for a 100-hour mission?
3. What mission time will result in a reliability of 0.95?

Answer 1

$$\lambda = \frac{1}{\theta} \quad \lambda = \frac{1}{5000} \quad \lambda = 0.0002 \text{ failures/hour}$$

Answer 2
The system reliability for a 100-hour mission is

$$R(t) = e^{-\lambda t}$$

$$-\lambda t = -0.0002 * 100 = -0.02$$

$$R(t) = e^{-0.02}$$
$$= \text{System reliability for a 100-hour mission}$$
$$\therefore R(t) = 0.98$$

Answer 3

$$R(t) = e^{-\lambda t}$$
$$0.95 = e^{-0.0002 * t}$$
$$\text{In } (0.95) = t * 0.0002$$
$$0.513 = t * 0.0002$$
$$t = 256.5$$

A mission time of 256.5 hours will result in a reliability of 0.95.

Region of Accelerated Failure and Wear-Out Failure

The is the wear-out period, and it has a normal distribution.

The right side of the bathtub curve has an increasing failure rate, signifying that

- Probability of failure is increasing
- Age is a factor in the probability of failure

This makes sense, as most items become more likely to fail with age due to accumulated wear.

The normal distribution can be used to model a product's times to failure in the wear-out period. This is because the normal distribution has an increasing failure rate, corresponding with that of the wear-out period. Both are shown in Figure 9.6.

To calculate reliability using the normal distribution, we must know the mean and the standard deviation:

$$z = \frac{t_1 - \mu}{\sigma} \tag{9.4}$$

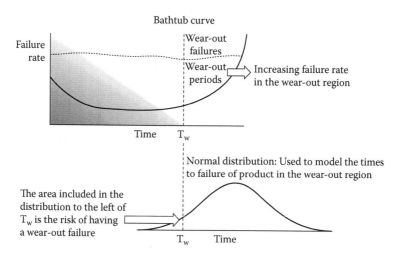

FIGURE 9.6
Wear-out period normal distribution.

where t_1 is the specific time and μ is the mean of the distribution, referred to as the mean wear-out time. When the normal distribution is used, 50% of the population will fail by the time the accumulated age of the product is equal to the mean wear-out time. σ is the standard deviation.

As an example, a compressor used in an industrial process has a normally distributed wear-out failure time. The compressor has a mean wear-out time of 8000 hours with a standard deviation of 760 hours.

1. What is the reliability $[R(t_1)]$ of the compressor for 7050 hours (or the probability that the compressor will not fail before it has accumulated 7050 hours of use)?
2. When should the compressor be replaced so that the probability of failure never exceeds 0.95?
3. If the compressor has accumulated 6500 hours of use without failing, what is the reliability for the next 100 hours (or, put differently, the probability that it will not fail in the next 100 hours)?

Answer 1

$$z = \frac{t_1 - \mu}{\sigma}$$

$$z = \frac{(7050 - 8000)}{760}$$

$$z = -1.25$$

In the Z table, reliability can be found against this value, -1.25. In the tables, it is the area to the right of the z values:

$$R(t_1) = 0.894$$

where 0.894 is the probability of the compressor not failing before accumulating 7050 operating hours. The probability (P) of the compressor failing in that period could also be calculated:

$$P(\text{system failure}) = 1 - R(t_1)$$
$$P(\text{system failure}) = 1 - 0.894 = 0.106$$
$$P(\text{system failure}) = 0.106$$

Answer 2

To determine when the compressor should be replaced, we must rearrange the translation equation to calculate for t_1:

$$t_1 = \mu + z(\sigma)$$

Since μ and σ are given, z is the only figure needed to find t_1. The reliability engineer must look up 0.95 (the target reliability figure) in normal probability tables to find that the required z is –1.65:

$$t_1 = \mu + z(\sigma)$$
$$t_1 = 8000 + (-1.65)(760)$$
$$t_1 = 8000 - 1254$$
$$t_1 = 6746 \text{ hours}$$

If the compressor is replaced after accumulating 6746 hours of use, the reliability (the probability of not failing) will never be less than 0.95.

Answer 3

To calculate reliability after a unit has successfully reached the age of t_1, it is necessary to use the formula

$$R(t/t_1) = R(t + t_1)/R(t_1) \qquad (9.5)$$

In this case, $t = 100$ and $t_1 = 6500$.

Here we need to calculate two reliabilities first: $R(6500)$ and $R(6500 + 100 = 6600)$

The Z value for $R(6500)$ is

$$Z = (6500 - 8000)/760$$
$$Z = -1.97$$

Following answer 1 logic,

$$R(t) = R(6500) = 0.9758$$

and the Z value for $R(6600)$ is

$$Z = (6600 - 8000)/760$$
$$Z = -1.84$$

Following answer 1 logic,

$$R(t + t_1) = R(6600) = 0.9673$$

Using Equation 9.5,

$$R(t/t_1) = R(t + t_1)/R(t_1)$$
$$R(t/t_1) = 0.9673/0.9758$$
$$= 0.9913$$

If the compressor has successfully accumulated 6500 hours, 0.9913 is the reliability for the next 100 hours.

Tolerance

Tolerance is a permissible limit of variation in a parameter's dimension or value.

Dimensions and parameters may vary within certain limits without significantly affecting the equipment's function.

Designers specify tolerances with a target and specification limits (upper and lower) to meet customer requirements (Figure 9.7).

The tolerance range, the difference between those limits, is the permissible limit of variation.

Statistical Tolerance

Parts work together, fit into one another, interact together, and bond together.

Since each part has its own tolerance, statistical tolerance is a way to determine the tolerance of an assembly of parts.

By using sample data from the process, statistical tolerance defines the amount of variance in the process. Statistical tolerance is based on the

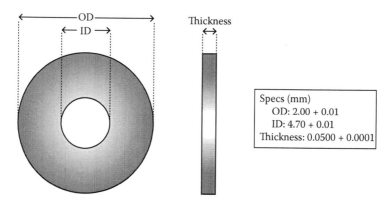

FIGURE 9.7
Part dimensional specification.

relationship between the variances of independent causes and the variance of the overall results:

$$X \pm Ks$$

where X is the sample mean, s is the standard deviation, and K is the constant from a two-sided tolerance table.

For example, given a 12-piece sample from a process with a mean of 14.591 and a standard deviation of 0.025, find the tolerance interval so that there is 0.95 confidence that it will contain 99% of the population.

From Table 9.1, $K = 4.150$.

$$\text{X-bar} \pm Ks = 14.591 \pm (4.150 \times 0.025)$$

This gives us

$$14.591 + 0.104 \quad \text{and} \quad 14.591 - 0.104$$

The tolerance interval becomes

$$14.695/14.487$$

Stack Tolerance

Sometimes parts are stacked together (Figure 9.8). Depending on the application, the parts may be the same or quite different. In these cases, tolerance levels must be determined for the entire stack.

TABLE 9.1

Factors for Tolerance Intervals Values of K for Two-Sided Intervals

Confidence Level	0.90			0.95			0.99		
Percent Coverage	0.90	0.95	0.99	0.90	0.95	0.99	0.90	0.95	0.99
2	15.978	18.800	24.167	32.019	37.674	48.430	160.193	188.491	242.300
3	5.847	6.919	8.974	8.380	9.916	12.861	18.930	22.401	29.055
4	4.166	4.943	6.440	5.369	6.370	8.299	9.398	11.150	14.527
5	3.949	4.152	5.423	4.275	5.079	6.634	6.612	7.855	10.260
6	3.131	3.723	4.870	3.712	4.414	5.775	5.337	6.345	8.301
7	2.902	3.452	4.521	3.369	4.007	5.248	4.613	5.488	7.187
8	2.743	3.264	4.278	3.136	3.732	4.891	4.147	4.936	6.468
9	2.626	3.125	4.098	2.967	3.532	4.631	3.822	4.550	5.966
10	2.535	3.018	3.959	2.839	3.379	4.433	3.582	4.265	5.594
11	2.463	2.933	3.849	2.737	3.259	4.277	3.397	4.045	5.308
12	2.404	2.863	3.758	2.655	3.162	**4.150**	3.250	3.870	5.079

Source: http://www.bessegato.com.br/UFJF/resources/tolerance_table.pdf.

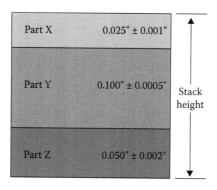

FIGURE 9.8
Stack tolerance.

Assuming the processes producing each part are capable and within normal distribution, the tolerances of the parts are not additive, but are instead related to the variance:

$$\text{Tolerance} = \sqrt{T_{\text{Part X}}^2 + T_{\text{Part Y}}^2 + T_{\text{Part Z}}^2}$$

$$\text{Tolerance} = \sqrt{0.001 + 0.0005 + 0.002}$$

$$\text{Tolerance} = \sqrt{0.0000052}$$

$$\text{Tolerance} = 0.0023$$

(9.6)

Statistical Tolerancing

Tolerances are usually set without knowing which manufacturing process will be used to manufacture the part, so the actual variances are not known.

However, a worst-case scenario would be where the process was just barely able to meet the engineering requirement. This situation occurs when the engineering tolerance is 6 standard deviations wide (±3 standard deviations). Thus, we can write the equation as

$$T \text{ stack} = T \text{ layer } 1 + T \text{ layer } 2 + T \text{ layer } 3$$

"In other words," Pyzdek (2003) asserts, "instead of simple addition of tolerances, the squares of the tolerances are added to determine the square of the tolerance for the overall result." Pyzdek goes on to say, "The result of the statistical approach is a dramatic increase in the allowable tolerances for the individual piece parts." This is an important concept in terms of tolerance because now the parts can have a greater tolerance for each part. An example is provided below (The Six Sigma Hand Book, Chapter 16: Statistical Tolerancing, pp. 600, 605).

Consider a shaft and bearing assembly where the shaft is specified to be 0.997 ± 0.001 and the bearing is specified to be 1.000 ± 0.0001. In this example, the minimum clearance between the two is 0.001 inch and the maximum clearance is 0.005 inch.

Thus, the assembly tolerance can be computed as

$$T_{assembly} = 0.005 - 0.001 = 0.004$$

The statistical tolerancing approach is used here in the same way as it was used above. Namely,

$$T_{assembly} = 0.004 = \sqrt{T_{bearing}^2 + T_{shaft}^2} \qquad (9.7)$$

If we assume equal tolerances for the bearing and the shaft, the tolerance for each becomes

$$T_{assembly} = \sqrt{(0.004)^2 / 2} = \pm 0.0028$$

which nearly triples the tolerance for each part.

Marketing and Porter's Five Forces Analysis

Marketing

In order to understand good business process, it is necessary to understand marketing and selling processes.

Marketing is the management process through which goods and services move from concept to the customer. It includes the coordination of four elements called the four P's of marketing:

1. Identification, selection, and development of a *product*
2. Determination of its *price*
3. Selection of a distribution channel to reach the customer's *place*
4. Development and implementation of a *promotional* strategy

For example, new Apple products are developed to include improved applications and systems, are set at different prices depending on how much capability the customer desires, and are sold in places where other Apple products are sold. In order to promote its devices, the company features its debuts at tech events and highly advertises them on the web and on television.

Marketing is based on thinking about the business in terms of customer needs and their satisfaction.

Marketing differs from selling because, in the words of Harvard Business School's retired professor of marketing Theodore C. Levitt, "Selling concerns itself with the tricks and techniques of getting people to exchange their cash for your product. It is not concerned with the values that the exchange is all about. And it does not, as marketing invariably does, view the entire business process as consisting of a tightly integrated effort to discover, create, arouse and satisfy customer needs." In other words, marketing has less to do with getting customers to pay for your product and more to do with developing a demand for that product and fulfilling the customer's needs (Levitt 2006).

Porter's Five Forces Analysis

Five forces analysis assumes that there are five important influences on a company that exert forces on its ability to serve its customers and make a profit (Figure 9.9).

In 1979, Michael Porter listed five forces that affect the success of an enterprise (Porter 1980):

- Bargaining power of customers: This force represents the ability and buying power of your customers to drive prices down.
- Bargaining power of suppliers: This force represents the ability and power of your suppliers to drive up prices of their products.
- Threat of new entrants: This force represents the ease by which new competitors can enter the market and drive prices down.
- Threat of substitute products: This force represents the extent to which different products or services can be substituted for your own.
- Intensity of competitive rivalry: This force represents the strength of the competition in your industry.

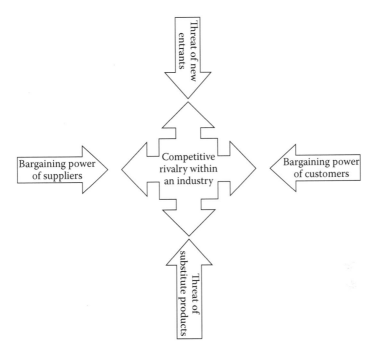

FIGURE 9.9
Porter's five forces analysis.

There are three key steps to deal with driving forces:

Step 1: Identify forces likely to exert the greatest influence over the next one to three years. Usually no more than three or four factors qualify as real drivers of change.

Step 2: Assess impact. Are the driving forces acting to cause market demand for product to increase or decrease? Are the driving forces acting to make competition more or less intense?

Will the driving forces lead to higher or lower industry profitability?

Step 3: Determine what strategy changes are needed to prepare for impacts of driving forces.

TRIZ

TRIZ is a relatively new approach to enhancing robust design. TRIZ aids in the decision-making and problem-solving processes during the design phase of DMADV. TRIZ is a way of lateral thinking. This innovative technique is based on two principles:

1. Somebody, sometime, somewhere has already solved your problem or one similar to it. Creativity means finding that solution and adapting it to the current problem.
2. Don't accept defects or deviations. Resolve them. Use them as a resource to solve problems.

TRIZ is an acronym for the Russian phrase *Teorija Rezbenija Izobretaltelshih Zadach*, meaning "theory of inventive problem solving."

Genrich Altshuller (1926–1998), a Russian mechanical engineer, created TRIZ as a set of problem-solving design tools and techniques. After studying more than 400,000 patents looking for inventive problem-solving methods, Altshuller noticed patterns across different industries.

Traditionally, inventive problem solving is linked to psychology; however, TRIZ is based on a systematic view of the technological world. Altshuller realized that people, including specialists, have difficulty thinking outside of their field of reference. Given a problem (P) within their specialty, many people will only limit their search for a solution (S) to their area of specialty (Figure 9.10).

What happens if the known solution to the problem could be found in another knowledge area? Below is an example of a solution found in a seemingly unrelated area.

Traditional diamond-cutting methods cut diamonds along natural fractures, but often result in new fractures that go undetected until using the diamond. Rather than improving the existing process, cutters needed a new process. The key to establishing the new method was a pressurized process in the food canning industry used to split green peppers and remove the seeds. A similar technique is also used in cutting

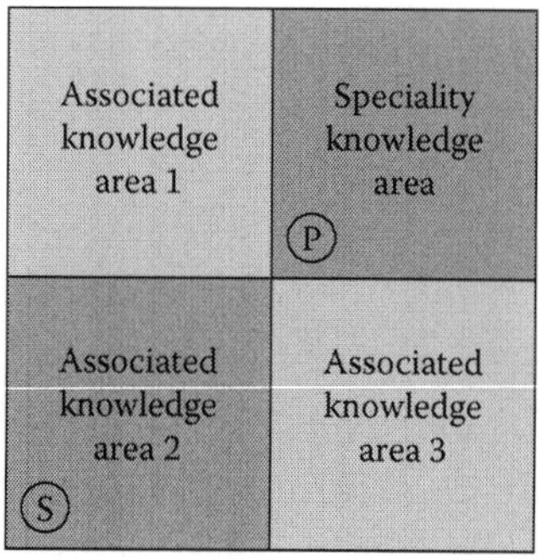

FIGURE 9.10
Problem of psychological inertia.

TABLE 9.2
Some Links to TRIZ and DFSS

Stage	Breakthrough Strategy Phase	Objective
Identifications	Recognize	Identify key business
	Define	The problem and customer response
Characterization	Measure	Understand current performance levels
	Analyze	
Institutionalization	Standardize	Transform how day-to-day business is conducted
	Integrate	

oversize broccoli in the canning industry. Water jet cutting applied to diamond cutting resulted in cuts without additional damage to precious diamonds.

TRIZ and DFSS

By providing a methodology to think and look outside the box and avoiding contradictions, TRIZ can help engineers, designers, developers, researchers, and quality professionals solve problems and find new ideas leading to new product development (Table 9.2).

Some common traits of TRIZ, DFSS, and Six Sigma include

- Solving bottlenecks
- Eliminating contradictions discovered in the house of quality roof
- Determining target values
- Identifying potential failure modes
- Lowering costs
- Improving serviceability

10

Case Study: Improvement Using Lean Six Sigma

Cricket was once known as a gentleman's game, the sport for the bourgeoisie, for cultured men. Modern cricket, however, is all about cutthroat competition and dogfights. On-field altercations, match fixing, and outrageous fan behavior have made the game not so gentlemanly.

Long gone are the days when every beautiful shot was greeted with cultured applause at Wankhede's cricket stadium in Mumbai. The muffled sound of clapping has now been replaced by throaty abuse, boos, and hooting with accompaniments like drums and cymbals.

In such chaotic conditions, mastery in the game is a must to get selected in the competitive matches. Here Lean Six Sigma (LSS) helps. Below is a case study of a bowler-turned-batsman who not only improved his batting, but also regained his social standing.

Here we are at the cricket ground in Mohali, Punjab state, India, where we will see how the LSS methodology can be put to use to improve batting in the game of cricket of one Mr. Singh.

Table 10.1 shows the Six Sigma road map.

RECOGNIZE AND IDENTIFY KEY BUSINESS: GAME OF CRICKET

Cricket is a popular sport in England, Australia, the Indian subcontinent (India, Pakistan, Sri Lanka, and Bangladesh), South Africa, New Zealand, the West Indies, and a few other countries, such as Afghanistan, Ireland, Kenya, Scotland, the Netherlands, and Zimbabwe (Figure 10.1).

TABLE 10.1

R-DMAIC-SI Strategy

Stage	Breakthrough Strategy Phase	Objective
Identifications	Recognize Define	Identify key business the problem and customer response
Characterization	Measure Analyze	Understand current performance levels
Institutionalization	Standardize Integrate	Transform how day-to-day business is conducted

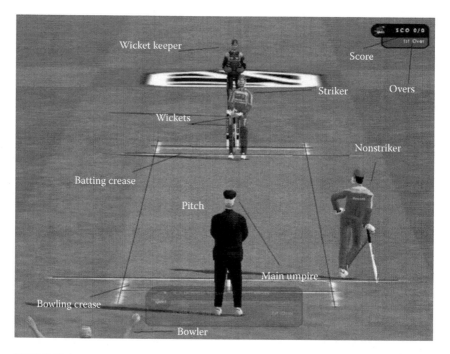

FIGURE 10.1
Game of cricket.

The game is played between two teams of 11 players each who score runs (points) by running between two sets of three wooden stumps called wickets. Each of the wickets is at one end of a rectangular-shaped field called the pitch. Around the pitch is a much larger oval of grass called the cricket ground (Figures 10.2 through 10.4).

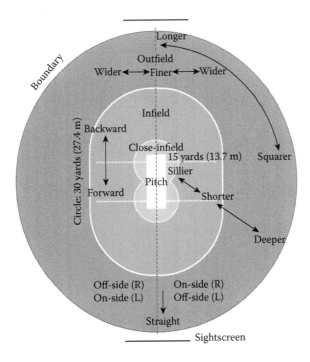

FIGURE 10.2
Oval-shaped cricket ground with pitch.

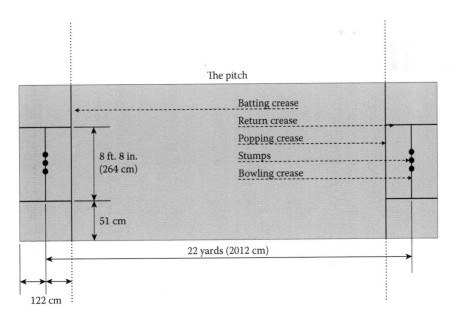

FIGURE 10.3
Dimensions of the pitch.

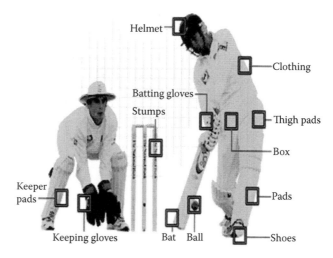

FIGURE 10.4

Batsman and wicket keeper with personal protection equipment.

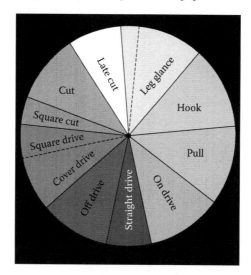

FIGURE 10.5

Names of the directions in which a ball can be hit.

Figure 10.5 shows the directions in which a right-handed batsman may send the ball when playing various cricketing shots. The diagram for a left-handed batsman is a mirror image of this one. Table 10.2 shows the techniques of facing a ball that a batsman should learn.

The basic batting statistics of our interest include

- Batting average (ave): The total number of runs divided by the total number of innings in which the batsman was out:

TABLE 10.2

Techniques of Facing a Ball That a Batsman Should Learn Well

Length/Line	Outside of Stump	Off Stump	Straight	Leg Stump	Outside Leg Stump
Half volley	Front foot Cover drive	Front foot Off drive	Front foot Straight drive	Front foot On drive	Front foot Flick off the legs
Good length	Leave	Front foot Defense	Front foot Defense	Front foot Defense	Front foot Leg defense
Short of length	Leave	Back foot Defense	Back foot Defense	Back foot Defense	Back foot Leg defense
Short	Back foot Cover drive	Back foot Off drive	Back foot Straight drive	Back foot On drive	Back foot Leg defense
Long hop	Cut	Pull	Pull	Pull	Pull

$$\text{Batting average} = \frac{\text{Runs}}{[\text{Innings} - \text{NO (not out)}]}$$

- Strike rate (SR): Number of runs scored per 100 balls faced (BF):

$$SR = \frac{[100 * \text{Runs}]}{BF}$$

- Run rate (RR): Number of runs a batsman (or the batting side) scores in six balls

CASE STUDY LEARNING OBJECTIVES

Upon completing the case study you will be able to

- Identify evidence of Singh's cricketing problem
- Identify the impact Singh's problems are having on his quality of life
- Explain Singh's investment in cricket
- Describe how Singh and Gary, the coach, understand the problem
- Describe various measurements and analysis tools Gary and Singh use in their effort to utilize Lean Six Sigma principles to help Singh improve his game

- Identify measures and metrics used to analyze specific aspects of Singh's game
- List the root causes that have most impacted Singh's cricket performance
- Identify three key action steps (vital few) Singh is taking to improve his game
- Describe Gary and Singh's plan to ensure Singh will continue to improve his game

---

DEFINE PHASE: IDENTIFY EVIDENCE OF SINGH'S CRICKETING PROBLEM

A problem well defined is half solved.

Here Gary helps Singh define the processes and importance of

$$Y = f(X)$$

The process outcome Y is a result (function) of process inputs X.

In the define phase, we understand the project outputs (Ys) and how to measure them.

Here voice of the customer (VOC) and SIPOC are needed.

SIPOC provides big Ys and little Ys. It also provides Xs.

To recap SIPOC,

- *Suppliers* provide inputs to the process.
- *Inputs* define the material, service, and information that are used by the process to produce the outputs.
- *Process* is a defined sequence of activities, and it usually adds value to inputs to produce outputs for the customers.
- *Outputs* are the products, services, and information that are valuable to the customers.
- *Customers* are the users of the outputs produced by the process.

10 Steps Leading to the Team Charter

D1: Singh's investment
- Budding cricketer
- Practices twice a week

- Drums up business contacts
- Lucrative deals begun on the cricket ground
- Practicing exercises to improve health
- Cricketing game not improved over last five years
- Scores strike rate 33, batting average 5
- Friends and business partners 50% better

D2: Gary's credentials

- Singh has asked his cricketing buddy Gary to help him improve his game.
- Gary will coach him on watching the ball from the bowler's hand in the run-up all the way to the bat, so one can determine which way the bowlers may be trying to swing the ball. Focus on the amount of swing.
- Most experienced cricketers have other areas that can produce significant results, such as foot work, various shots, and stance.
- Gary is an excellent cricketer and expert in LSS.

D3: Singh's current state

- Singh was investing a lot of time and money.
- His performance was not improving.
- This leaves him on the sideline when his business contacts and friends are playing.
- He is frustrated and has lost business opportunities.
- He planned on cricketing the rest of his life.
- He is motivated to improve his cricket game.

D4: What is Singh's problem?

- His business interests hamper his game.
- Friends and contacts are better players.
- He doesn't practice enough.
- His performance has not improved in five years.
- His scores from five years ago, over 15 one-day internationals (ODIs), are
  - Strike rate (runs per 100 balls) = 16
  - Batting average (number of runs scored/number of times out) = 5
- Recent scores over 15 ODIs
  - Strike rate 17
  - Batting average 5

D7: Impact of Singh's problem?

- Three business contacts have distanced themselves.
- They score better and play in better matches.
- His best friend has almost disappeared.
- His wife, who plays on a women's team, has a better strike rate.

D8: Scorecard five years ago

Why do scorecards from five years ago provide evidence of Singh's problem?

- Demonstrate that Singh is not playing enough cricket
- Demonstrate a trend that shows that Singh has not improved in five years
- Demonstrate that Singh's playing skills are out of date
- Evidence that Singh is using antiquated equipment

D9: All of the following impact Singh

- His wife is a better cricketer.
- He sees less and less of his best friend.
- He is losing his business contacts.

D10: Problem: Cricket score is too low

$$Y = f(X)$$

Big Ys (high-level, high-impact metrics)

- Cricket score
- Retirement options
- Social options
- Revenue
- Health and fitness
- Quality of life

Little Ys (linked to vital X)

- Average score per match
- Strike rate per match
- Batting average
- Hours played per match
- Business contacts made
- Match outings planned

Vital X

- Batting proficiency (key performance)
- Core processes (performance drivers of vital X)

Practice Drill Process

- Warm-up process
- Physical training process
- Mental training process
- Equipment selection process
- Batting process
- Facing spin bowling process
- Facing fast bowling process
- Field placement strategy process

Gary and Singh's Team Charter

See Figure 10.6.

Business Case
Focuses on Big Ys of the game that impact Singh's personal and business life.
Improvement in his batting skills will improve his health, his relationships, marriage and business opportunities.

Opportunity statement
Achievement of the goal consistently will imporeve Singh's mental and physical health and would imporve both social and business opportunities. Singh will be able to improve his business contact with a projected increase of 50% in his imcome.

Goal Statement
To improve Singh's rate and batting score by 50%

Project Scope
Include all preparation, practice and playing of Cricket at the Mohali cricket ground between 1/5 and 1/19.

Project Plan (DAIMC Process)
1/5 to 1/6--- Define Problem, Identify Measurement system
1/6 to 1/7--- Run tests and gather data
1/7 to 1/8--- Analyze data and identify root causes
1/8 to 1/9--- Identify and test solutions
1/9 to 1/10- Implement solutions and control performance

Team Members:
Sponser: Singh's wife Tara
Process Owner: Singh
Black Belt: Gary
Process Expert: Dutta-Ground Expert
Process Expert: Shastri-Batting Expert
Process Expert: Milkha-Physical Trainer
Process Expert: Srivastav-Psychiatrist

FIGURE 10.6
Team charter.

MEASURE PHASE

The strategic importance of measurement is that

1. Strategic importance is well understood.
2. Everyone understands the improvement opportunity.

In this phase we need to measure the performance. Here the priority is to improve the quality of the game.

In terms of cricket, this means measure the

- Number of "dot" balls
- Number of singles
- Number of boundaries
- Number of sixes
- Number of times for getting out

Critical to Process

This is how the process is performing in real time.

Singh and Gary must measure factors that drive performance.

The possible critical to processes (CTPs) are

- Strike rate
- Batting average

Baseline data from the previous five test matches last year are

- Strike rate: 17
- Batting average: 5

Table 10.3 shows the measurement plan (10 overs fast and 10 overs spin).

ANALYZE PHASE

See Figures 10.7 and 10.8.

TABLE 10.3

Measurement Data Sheet

Measure	Definition of Measure	Data Source and Location	Sample Size	Who Will Collect	How Collected	When Collected	Gary's Observations
Dot balls					On scorecard mark yes or no		Notes
Singles					Count		Notes
Fours					Count		Notes
Sixes					Count		Notes
Out					Count		Notes

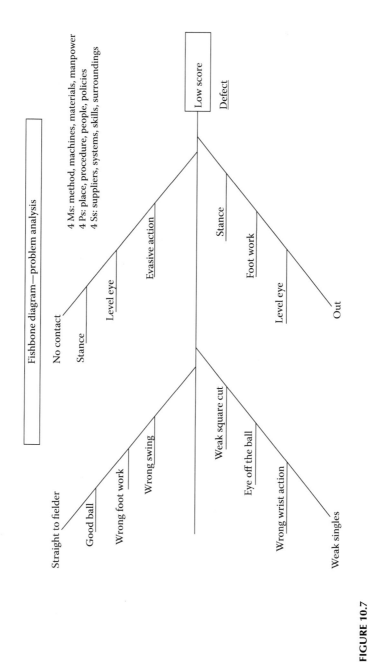

FIGURE 10.7
Problem analysis.

Problem	Stroke Count	% of Total	Cumm. %
Stance	80	39	39
Front foot work	50	25	64
Back foot work	40	20	84
Wrong square cut	20	10	94
Grip	10	5	99
Evasive action	4	1	100

(a)

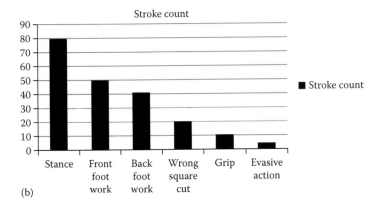

(b)

FIGURE 10.8

Data analysis using Pareto chart. (a) Record of problem data for analysis phase and (b) problem data Pareto chart for analysis phase.

IMPROVE PHASE

Singh and Gary examined the root cause analysis and Pareto chart data. They found that the following were the main contributors in Singh's substandard batting performance:

- Less efficient stance
- Front foot work
- Back foot work
- Square cut techniques

They came up with an action plan, shown in Table 10.4.

Singh's Practice and Training Regime

Singh and Gary came up with a training regime that focuses on four solutions. It keeps track of what he learned and his results and progress.

TABLE 10.4

Action Plan

Solution	Action
Improve stance, grip, and back lift	Improve stance, grip, and back lift
Front foot work	Modify physical exercises
Back foot work	Buy professional "stance" improvement books and videos
Square cut	Practice all positions while batting one hour per week
Learn and practice stress reduction and visualization	Meeting with sports psychologist
Be ready to concentrate on the next shot	Learn emotional management technique, especially after poor shot

Singh and Gary would review the data and verify that the four solutions were implemented consistently and producing the desired results.

CONTROL PHASE

Having completed an improvement program, Singh is now ready to put his new batting practice into action.

He faced 100 balls each from the fast and spin bowlers once a week for four weeks. All data were tracked and monitored.

Three outcomes resulted:

1. Improve stance, grip, and back lift
   - Average improvement of 10 in strike rate
   - Reduction in number of times out—thus average improvement of batting rate by 3
2. Improve front foot work
   - Average improvement of 15 in strike rate
   - Reduction in number of times out—thus average improvement of batting rate by 4
3. Mental lapses after poor shots
   - All eliminated
   - Average improvement of 18 in strike rate

STANDARDIZE

- The approach is redefined through regular cycles of improvement twice per year. The improvements come through corrective actions based on team comments and expectations, coach's inputs for improvement, and Lean Kaizen efforts.
- All these changes are recorded, documented, and communicated to all stakeholders through changed techniques, work instructions, and equipment changes as required.

INTEGRATE

- This process approach is aligned with other team project requirements like strategic planning by defining process metrics and monitoring their effectiveness and efficiency.
- The approach is aligned with stakeholders' needs.
- It forms the backbone of the plan–do–check–act (PDCA) for an overall team leadership system.
- It integrates with the Global Quality Management System (GQMS) and measurement and analysis systems as the execution arm to "drive" the game performance to next year's goals.
- The process synchronizes with reward and recognition systems to drive passionate work culture, heavily integrated into goal setting for a visual and organized approach to a high-performance team management system.

Table 10.5 shows the continuous improvement data.

TABLE 10.5

Continuous Improvement Data

	Matches	Innings	Runs	Not Out	Average	Strike Rate	100s	50s
Test	34	52	1639	6	35.63	58.91	3	9
ODI	274	252	8051	38	37.62	87.58	13	49
T20	23	22	567	4	31.5	151.60	0	5
IPL	44	43	915	6	24.72	132.03	0	3

Glossary

5S: Stands for five Japanese words that begin with letter S. The words as translated in English are *seiri* (*sort*), *seiton* (*straighten* or *set in order*), *seso* (*shine*), *seiketsu* (*standardize*), and *shitsuke* (*sustain*). All together, they mean orderly, well-organized, well-inspected, clean, and efficient workplaces.

5 whys: Simple process of determining the root cause of a problem by asking why after each situation to dive deeper, in more detail, to arrive at the root cause of an issue.

7 wastes: Originally identified by Taiichi Ohno, these are (1) overproduction, (2) waiting, (3) transportation, (4) overprocessing, (5) stock on hand, (6) movement, and (7) making defective product.

8D: Popular method for problem solving because it is reasonably easy to teach and effective. The 8D steps and tools used are as follows:

> **D0:** Preparation for the 8D
>
> **D1:** Formation of a team
>
> **D2:** Description of the problem
>
> **D3:** Interim containment action
>
> **D4:** Root cause analysis (RCA) and escape point
>
> **D5:** Permanent corrective action
>
> **D6:** Implementation and validation
>
> **D7:** Prevention
>
> **D8:** Closure and team celebration

This process is known as Global 8D by Ford.

A3: Report prepared on 11 × 17 inch plain paper by the owner of the issue. The PDCA format is used. It gathers current information and its analysis, creates goals and metrics, and builds buy-in from stakeholders.

Andon: Japanese word meaning "light" or "lantern." It is a form of communication for an abnormal condition or machine malfunction. It often resembles a stop traffic light where red is stop, yellow is caution, and green is go. Another form can be an andon cord, which is pulled by the operator to communicate an abnormal situation.

AS 9100: Widely adopted and standardized quality management system for the aerospace industry.

BE: Business excellence.

Black Belt: Professional who can explain and practice Six Sigma philosophies and principles, including supporting systems and tools.

BPR: Business process reengineering. Analysis and redesign of workflow within and between enterprises. BPR reached its maximum popularity in the early 1990s.

budget: Estimate of costs, revenues, and resources over a specified period, reflecting a reading of future financial conditions and goals.

CA: Corrective action taken to eliminate the cause of nonconformity.

cause-and-effect diagram: This diagram-based technique helps us identify all of the likely causes of the problems faced in working environments.

changeover: Setting up a machine or production line to make a different part number or product.

changeover time: Time from the last good piece of the current production run to the first good piece of the next run.

constraint: Anything that limits a system from achieving higher performance. It is also called a bottleneck.

continual improvement: *Continual* indicates duration of improvement that continues over a long period of time, but with intervals of interruption, for example, a plant modification disrupted by logistics or traffic for nearly two years.

continuous improvement: Approach of making frequent and small changes to process whose cumulative results lead to higher levels of quality, cost, and efficiency.

countermeasure: Corrective action taken to address problems or abnormalities.

customer: Party that receives or consumes products (goods or services) and has the ability to choose between different products.

cycle: Sequence of operations repeated regularly.

cycle time: Time for one sequence of operations to occur.

effectiveness: Degree to which objectives are achieved and the extent to which targeted problems are solved. In contrast to efficiency, effectiveness is determined without reference to costs, and whereas efficiency means "doing the thing right," effectiveness means "doing the right thing."

EFQM: European Foundation for Quality Management.

equipment availability: Percentage of time equipment (or process) is available to run. This is sometimes called uptime.

error proofing: *See* Poka Yoke.

external setup: Procedures that can be performed while a machine is running.

FAI: First article inspection.

FIFO: First in, First out. In other words, material produced by one process is consumed in the same order (FIFO) by the next process.

fishbone diagram: Identifies many possible causes for an effect or problem. It can be used to structure a brainstorming session. It immediately sorts ideas into useful categories. Major categories of causes of the problem are

- Methods
- Machines (equipment)
- People (manpower)
- Materials
- Measurement
- Environment

flow: Completion of steps within a value stream so that product or service "flows" from the beginning of the value stream to the customer without waste.

flow production: Same as *flow.*

FMEA: Failure mode and effects analysis. A step-by-step approach for identifying all possible failures in a design, a manufacturing or assembly process, or a product or service. *Failure mode* means the way, or mode, in which something might fail. Failures are any errors or defects, especially ones that affect the customer, and they can be potential or actual. *Effects analysis* refers to studying the consequences of those failures.

FPY: First-pass yield. Number of units coming out of a process divided by the number of units going into that process over a specified period of time. Only good units with no rework are counted as coming out of an individual process. Also known as throughput yield (TPY).

Gemba: Japanese word meaning "real place," where action takes place—a shop floor or work areas.

Gemba walk: Walk carried out by a coach (a Lean Sensei) and student or students to look for abnormal conditions, waste, or opportunities for improvement.

Heijunka: Method for leveling production for mix and volume.

Hoshin Kanri: Strategic decision-making tool used for policy deployment.

internal setup: Procedures that must be performed while the machine is stopped.

Ishikawa diagram: *See* fishbone diagram.

Jidoka: Device that stops production or equipment when a defective condition arises. Attention is drawn to this condition and the operator who stopped the production. The Jidoka system has faith in the operator who is trained in the job.

just in time (JIT): Originally developed by the Toyota Production System (TPS). JIT presupposes that all waste is eliminated from the production line and only the inventory in the right quantity and at the right time is used for the production, where the rate of production is exactly as required by the customer.

Kaikaku: Japanese word meaning "innovation" or a "radical breakthrough." Kaikaku requires radical thinking and takes more time in planning and implementation.

Kaizen: Japanese word meaning "change for the better" or "do good." It is a process of making continual improvements by everyone keeping in mind quality and safety.

Kaizen event: Short team-based improvement project. Also called Kaizen blitz.

Kanban: Means "sign board" or a "label." It serves as an instruction for production and replenishment.

KCC: Key critical characteristic.

KPC: Key performance characteristic.

KPI: Key performance indicator.

lead time: Time required to move one piece from the time the order is taken until it is shipped to the customer.

line balancing: Technique where all operations are evenly balanced and staffing is also balanced to meet the takt time.

Manufacturing Resource Planning (MRP II): Like MRP, but takes into consideration the capacity planning and finance requirement. It works out alternative production plans through the simulation tool.

Materials Requirement Planning (MRP): Computerized system of determining quantity and timing requirements for production and delivery of products for customers as well as suppliers. This is a push production system.

MBNQA: Malcolm Baldrige National Quality Award. An award given to an organization for achieving the highest-quality standards.

milk run: Routing of supply and delivery trucks or vehicles to make multiple pickups and deliveries at various locations to reduce transportation waste.

Muda: Japanese word for waste. It is an element that does not add value to the product or service. Also known as no-value-added activity carried out on a product or service that does not add value and the customer will not pay for it.

Mura: Japanese word for variability or unevenness.

Muri: Japanese word for physical and mental strain or overburden.

One- (single) piece flow: Practiced in JIT system where one work piece flows from process to process to minimize waste.

operational excellence (Opex): Element of organizational initiative that stresses the application of a variety of principles, systems, and tools toward the sustainable improvement of key performance metrics. This philosophy is based on continuous improvement, such as a quality management system, Lean manufacturing, and Six Sigma. Operational excellence goes beyond the traditional methods of improvement and leads to a long-term change in organizational culture.

overall equipment effectiveness (OEE): Product of the following key measures:
1. Operational availability
2. Performance efficiency
3. First-pass yield quality

PAIP: Process for performance analysis and improvement.

PCP: Process control plan.

PDCA: Plan–do–check–act cycle for continual improvement.

PFMEA: Process failure mode and effects analysis.

point-of-use storage (POUS): Storing or keeping materials, tools, information, and items near to where they are used.

Poka Yoke: Also known as mistake proofing. *Poka* in Japanese means "inadvertent mistake," and *Yoke* means "prevention." These can be simple, low-cost devices or sophisticated electromechanical devices to prevent the production of defective product.

process: Sequence of interdependent and linked procedures that, at every stage, consume one or more resources (employee time, energy, machines, and money) to convert inputs (data, material, parts, etc.) into outputs. These outputs then serve as inputs for the next stage until a known goal or end result is reached.

productivity: Measured as an output for a given input. Productivity increase is critical to improving living standards.

product realization: Term used to describe the work that the organization goes through to develop, manufacture, and deliver the finished product or service to the customer.

pull: Alternatively known as pull production, where the upstream supplier does not produce until the downstream customer signals the need.

push: Alternatively known as push production, where the upstream supplier produces as much as it can without regard to whether the downstream customer needs the product.

QOS: Quality operating system. Originally implemented by Ford. The methodology was established to measure the effectiveness of the quality system and to drive continuous improvement.

RPN: Risk priority number. In FMEA, RPN = Severity × Occurrence × Detection.

RTY: Rolled throughput yield. A probability that a single piece will pass through all production steps without a single defect.

shadow board: Board where each tool has a place, showing which tools are missing.

single-minute exchange of dies (SMED): Group of techniques developed by Shigeo Shingo for changeover of production equipment in less than 10 minutes.

SIPOC: Process identification where the requirements for the supplier, input, process steps, output, and customer are defined.

Six Sigma: Set of tools and techniques for process improvement. It was originally developed by Motorola in 1981.

spaghetti diagram: Diagram showing the layout and flow of information, material, and people in a work area. It is generally used to highlight motion and transportation waste.

SPC: Statistical process control. Quality control where process variations are measured and controlled.

standard work: Accurate description of every process step specifying takt time, cycle time, and minimum inventory needed and sequence of each process step. The entire process is carried out with minimum human motion and other wastes.

supermarket: Where parts are stored before they go to the next operation. The parts are managed using minimum and maximum inventory levels.

sustainability: Continued development or growth, without significant deterioration of the environment and depletion of natural resources on which human well-being depends.

SWOT: Strength, weakness, opportunity, and threat. Strength and weakness analysis guides us to identify the positives and negatives inside an organization (SW). Opportunity and threat analysis guides us to identify the positives and negatives outside of it. Developing a full SWOT analysis can help with strategic planning and decision making.

system: A group of interdependent processes and people that together perform a common goal.

takt time: Available production time divided by the rate of the customer demand.

TPM: Total productive maintenance. System to ensure that every production process machine is able to perform its required tasks such that production is not interrupted.

TQM: Total quality management. Management approach that originated in the 1950s. The TQM culture requires quality in all aspects of the company's operations, with processes being done right the first time and defects and waste eradicated from operations. To be successful implementing TQM, an organization must concentrate on eight key elements:

1. Ethics
2. Integrity
3. Trust
4. Training
5. Teamwork
6. Leadership
7. Recognition
8. Communication

VAA: Value-added activity.

value: Capability provided to a customer at the right time at an appropriate price. Defined by the customer.

value stream: Sequence of actions required to design, produce, and provide a specific good or service, and along which information, materials, and worth flow.

visual factory: Term to describe how data and information are conveyed to a Lean manufacturing environment. Here time and resources dedicated to conveying information are a form of waste. By using visual methods, information is easily accessible to those who need it. Visual information makes the current status of all processes immediately apparent.

work in process (WIP): Incomplete product or services that are waiting further processing.

Bibliography

Akao, Y., ed. *Quality Function Deployment*. Cambridge MA: Productivity Press, 1990.

American Society for Quality (ASQ). *The Certified Manager of Quality/Organizational Excellence*. Foundations in Quality Learning Series. Milwaukee, WI: ASQ Quality Press, 1994.

American Society for Quality (ASQ). *The Certified Quality Engineer*. Milwaukee, WI: ASQ Quality Press, 2000.

ASQ Lean Six Sigma Conference, Phoenix, AZ, March 8–9, 2010.

Automotive Industry Action Group (AIAG). *PFMEA*. 3rd ed. Southfield, MI: AIAG, 2001.

Automotive Industry Action Group (AIAG). *MSA*. 3rd ed. Southfield, MI: AIAG, 2002.

Benbow, Donald W., and Kubiak, T.M. *The Certified Six Sigma Black Belt Handbook*. Milwaukee, WI: ASQ Quality Press, 2005.

Brassard, Michael. *The Memory Jogger Plus+®: Featuring the Seven Management and Certified Six Sigma Black Belt Certification Preparations*. Salem, NH: Goal/QPC, 1996.

Crosby, Philip. Crosby's 14 steps to improvement, *Quality Progress*, December 2005: 60–64.

Crosby, Philip. *Quality without Tears*. New York: McGraw-Hill, 1984.

Deming, W. Edwards. *Out of the Crisis*. Cambridge, MA: MIT Press, 1986.

Deming, W. Edwards. *The New Economics for Industry, Government, Education*. 2nd ed. Cambridge, MA: MIT Press, 1993.

Dittmer, H. William. *Goldratt's Theory of Constraints: A Systems Approach to Continuous Improvement*. Milwaukee, WI: ASQ Quality Press, 1997.

Feigenbaum, Armand V. *Total Quality Control*. 3rd ed. New York: McGraw-Hill, 1991.

Galsworth, Gwendolyn. Visual-Lean Institute, Portland, OR: Visual Thinking Inc., 1992.

International Organization for Standardization (ISO). Quality management principles 2. Geneva: ISO, 2001.

International Organization for Standardization (ISO). ISO for small and medium-sized businesses: ISO/TC advice. Milwaukee, WI: ISO, 2008.

ISO 9001:2008. Quality management systems—Requirements. Geneva: International Organization for Standardization, 2008.

ISO 9004. Managing for the sustained success of an organization—A quality management approach. Geneva: International Organization for Standardization, 2010.

ISO/TR 10013:2001. Guidelines for quality management system documentation. Geneva: International Organization for Standardization, 2001.

ISO/TS 16949. Southfield, MI: Automotive Industry Action Group, 2009.

Keller, Paul. *Six Sigma Demystified*. New York: McGraw-Hill, 2005.

Krueger, Kaye. Value added vs non value added activities. October 15, 2015. Available at https://www.wisc-online.com/value-added-vs-non-value-added-activities.

Levitt, Theodore C. *Ted Levitt on Marketing*. Watertown, MA: Harvard Business Review Book. Chapter on "Selling versus Marketing" pp. 152–154, 2006.

Lowe, A.J. and Ridgway, K. Quality function deployment. Sheffield, UK: University of Sheffield, 2001. http://www.shef.ac.uk/~ibberson/qfd.html.

Maasaki, Imai. *Gemba Kaizen*. New York: McGraw-Hill, 1997.

Porter, Michael E. *Competitive Strategy*. New York: Free Press, 1980.

Pyzdek, Thomas. *The Six Sigma Handbook: A Complete Guide for Green Belts, Black Belts, and Managers at All Levels*. 2nd ed. New York: McGraw-Hill, 2003.

Pyzdek, T. and Keller, P.A. Project scheduling, *The Six Sigma Handbook*, 4th ed., p. 218, 2011.

Pyzdek, T. and Keller P.A. Quality Engineering Handbook. ISBN 0824746147.

Shiego, Shingo. *A Study of the Toyota Production System*, trans. Andrew Dillon. Boca Raton, FL: CRC Press, 2005.

SO/TR 10013:2001. Guidelines for quality management system documentation. Geneva: International Organization for Standardization, 2001.

Tague, N.R., *Quality Toolbox*, 2nd. ed., p. 558, ASQ Quality Press, March 30, 2005.

Toyota Production System. http://www.toyotageorgetown.com.

Wheeler, Donald J. *Advanced Topics in Statistical Process Control*. Knoxville, TN: SPC Press, 1995.

Womack, James P. and Jones, Daniel T. *Lean Thinking, Banish Waste and Create Wealth in Your Corporation*. Simon & Schuster, 2003.

Index

Page numbers ending in "f" refer to figures. Page numbers ending in "t" refer to tables.

A

A3 report, 203
Akao, Yoji, 40
Alternative hypothesis, 113; *see also*
    Hypothesis testing
Altshuller, Genrich, 184
Analysis of variance (ANOVA), 92, 107,
    110–111, 110t, 112t, 162
Analytical statistics, 7–8; *see also*
    Statistics
Analyze phase
    analysis of variance, 107, 110–111, 110t,
        112t, 162
    cause-and-effect, 107
    correlation tool, 108–109
    data classification, 108, 108f
    failure mode and effects analysis, 107,
        117
    fault tree analysis, 116–117, 117f
    five whys, 116–117
    gap analysis, 115–116
    goal performance, 115–116
    hypothesis testing, 112–115
    Pareto chart, 107
    performance analysis, 115–116
    regression analysis, 108–109
    relationships, 108–109, 108f, 109f
    root cause analysis, 116–117
    scatter diagrams, 108–109, 109f
    variables, 108–109, 108f, 109f
    waste analysis, 107
Andon cord, 203
Annual percentage rate (APR), 33–34
Appraiser variation (AV), 79–80
AS 9100, 74, 204
Attribute data charts, 140–153, 141f;
    see also Control charts

B

Balanced scorecard (BSC), 20, 21f, 22f
Bias, 77–78, 77f
Black Belt, 56, 164, 165, 204
Box and whisker plot, 53
"Breaking the Barriers" cartoon, 72f
Budget, 204
Burn-in, 170–171, 170f
Business excellence (BE), 1, 55t, 204
Business failure cost analysis, 27–28,
    28f, 29f
Business goals, 123
Business performance measures
    balanced scorecard, 20, 21f
    business failure cost analysis, 27–28,
        28f, 29f
    cost–benefit analysis, 30–32
    critical-to-quality measures, 21
    effectiveness of, 20f
    financial measures, 30–35
    key performance indicators, 20–21
    net present value, 32–35
    quality performance measures, 22–30
    return on assets, 32
    return on investment, 32
    stakeholders, 19–20
Business process improvement (BPI), 119
Business process reengineering (BPR), 204
Business processes
    balanced scorecard, 20, 21f, 22f
    customer strategies, 21f, 21t
    effectiveness of, 20f
    feedback loop, 20, 20f
    financial strategies, 21f, 21t
    internal business strategies, 21f, 21t
    learning and growth strategies, 21f, 21t
    stakeholders, 19–20

strategies, 20–21, 21f, 21t
team management, 35
Business strategies, 20–21, 21f, 21t

C

Cash flow (CF), 32–34, 34t
Cause-and-effect analysis, 107
Cause-and-effect diagram, 204
Cause-and-effect matrix, 119–120, 120f
Central limit theorem, 9–10, 9f
Central tendencies, 11–12, 12f
Changeover, 204
Changeover time, 204
Chart champion, 154
Chart location, 154
Chart types, 154; see also Control charts
Check sheets, 53
Circle diagram, 69, 70, 71f
Constraints, 69, 119–123, 204
Continual improvement, 204
Continuous flow manufacturing, 121
Continuous improvement, 34, 125, 158,
        201t, 204; see also Improve
        phase
Control chart constants, 82–83, 84t, 95,
        130, 132t
Control chart guide, 128–129, 129f
Control charts
    attribute data charts, 140–153, 141f
    c charts, 146–148, 147f–149f
    chart types, 154
    control chart constants, 82–83, 84t, 95,
        130, 132t
    control chart guide, 128–129, 129f
    control limits, 135f, 136f, 140f
    control phase and, 128–153
    control plan and, 154–155
    data charts, 130–140
    data sheets, 134f, 139f, 143f, 145f, 146,
        148f, 150f
    grand mean, 134f, 136f
    I-XR charts, 137–140, 138f–140f
    np charts, 143–146, 144f–146f
    p charts, 141–142, 142f–144f
    process stability and, 159
    s charts, 130–136, 133f–137f
    u charts, 146–147, 149–152, 150f–153f
    variable control charts, 130, 132f

variable data charts, 130–140
    X-bar charts, 130–137, 133f–137f
    X-MR charts, 137–140, 138f–140f
Control phase
    chart champion, 154
    chart location, 154
    chart types, 154
    control chart guide, 128–129, 129f
    control charts, 128–153
    control plan, 154–155
    Cp/Cpk and, 154
    critical-to-quality characteristic, 154
    documentation for, 157–158
    evaluations for, 158–159
    gage number, 154
    measurement method, 154
    measurement studies, 154
    process capability studies, 159
    process metrics, 157, 159
    process stability, 154, 159
    reaction plan, 154
    sampling plans, 154
    significant characteristic description,
        154
    significant characteristic number, 154
    standard operating procedures,
        157–158
    standard work instructions, 157–158
    statistical process control, 125–128,
        127f, 128f
    sustenance of improvements, 156–159
    take rate, 159
    training plan, 156–157
Control plan
    chart champion, 154
    chart location, 154
    chart types, 154
    Cp/Cpk and, 154
    critical-to-quality characteristic, 154
    example of, 154, 155f
    gage number, 154
    measurement method, 154
    measurement studies, 154
    process stability, 154, 159
    reaction plan, 154
    sampling plans, 154
    significant characteristic description, 154
    significant characteristic number, 154
    training plan, 156–157

Corrective action (CA), 201, 204
Correlation, 108–109, 109f
Correlation matrix, 41, 44, 48f
Cost metrics, 59–60
Cost of poor quality (COPQ), 24, 27f, 28
Cost–benefit analysis, 30–32
Countermeasure, 204
Cp/Cpk, 24, 25t–26t, 100–103, 154
Cricket case study, 187–201, 188f–190f;
    see also Lean Six Sigma
Critical path analysis (CPA), 62–63, 62f
Critical path method (CPM), 62–64,
    101, 104
Critical-to-cost (CTC) requirements, 51
Critical-to-delivery (CTD) requirements,
    51
Critical-to-process (CTP) requirements,
    51
Critical-to-quality (CTQ) flow-down, 21,
    37, 51–53, 52f
Critical-to-quality characteristic (CTQC),
    154
Critical-to-safety (CTS) requirements, 51
Critical-to-X (CTX) requirements, 51
Customer requirements, 43f, 50t, 119
Customer satisfaction, 38–39
Customer satisfaction survey, 39
Customer value, 37–39, 209
Customers
    competitor rating by, 43f
    concerns of, 168–171
    customer–supplier chain, 38, 39f
    definition of, 204
    eight hows, 45, 50t
    eight whats, 45, 50t
    external customers, 38
    five forces analysis and, 181–183, 183f
    internal customers, 38
    marketing and, 181–182
    needs of, 43f, 44f, 45f, 50t, 119
    reliability assurance for, 168–171
    requirements of, 43f, 50t, 119
    satisfaction of, 38–39
    satisfaction survey for, 39
    strategies for, 21f, 21t
    voice of, 39f
Customer–supplier chain, 38, 39f
Cycle time, 69, 204
Cycles, 204

D

Data, qualitative, 73
Data, quantitative, 73
Data charts, 130–153; *see also* Control
    charts
Data collection, 72–73
Data spread, 10
Defect assessment levels, 15–17, 15f,
    16f, 16t
Defects per million opportunities
    (DPMO), 13, 15, 16f, 16t, 23–24,
    25t–26t
Defects per unit (DPU), 22–30
Define, measure, analyze, design, and
    verify (DMADV), 164, 164f
Define, measure, analyze, design,
    optimize, and verify
    (DMADOV), 164–165
Define, measure, analyze, improve, and
    control (DMAIC) process, 1–5,
    5f, 7–8, 188t
Define phase
    business case, 54, 55t
    critical-to-quality flow-down, 37, 51–53
    goals, 54–55, 55t
    program evaluation and review
        technique, 37, 63–66
    project charter, 53–60, 58f
    project goal, 54–55, 55t
    project performance, 56, 57f, 58f, 59f
    project scope, 54
    project statement, 54–55, 56t
    project status report, 59f
    project tracking, 60–63
    quality function deployment, 37, 40–51
    SMART goals, 56
    tools for, 53
    voice of customer, 39f
Deming, W. Edwards, 7, 27, 125
Design for assembly (DFA), 167
Design for manufacturability (DFM),
    166–167
Design for producibility (DFP), 167
Design for Six Sigma (DFSS)
    define, measure, analyze, design,
        and verify, 164, 164f
    define, measure, analyze, design,
        optimize, and verify, 164–165

design for X, 166–168, 169f
diagram of, 162f
explanation of, 161–162
five forces analysis and, 181–183, 183f
marketing and, 181–182
methodologies for, 163–164
product life cycle cost, 163, 163f
psychological inertia and, 184, 184f
reliability and, 168–177
return on investment and, 162–163
teams for, 165–166
tolerance and, 177–181
tools for, 162–164
TRIZ approach and, 183–185, 185t
Design for X (DFX)
assembly and, 167
concepts of, 166–168, 169f
cost and, 166
manufacturability and, 166–167
producibility and, 167
safety and, 167–168
serviceability and, 167–168
techniques for, 166–168, 169f
tests and, 167
Design of experiments, 53
Discounted cash flow (DCF), 32–33
Dispersion, 12–13, 13f
DMAIC process, 1–5, 5f, 7–8, 188t

E

Effectiveness, 204
8D steps, 203
Eight hows/whats, 45, 50t
Equipment availability, 121, 205
Equipment variation (EV), 79–80
Error proofing, 121, 166, 168, 205
European Foundation for Quality
    Management (EFQM), 204
Expanded house of quality, 41–51,
    42f–49f
Experiment designs, 53
External setup, 205

F

Failure mode and effects analysis (FMEA),
    107, 117, 119, 205
Fault tree analysis (FTA), 116–117, 117f

Financial measures
    cash flow, 32–34, 34t
    cost–benefit analysis, 30–32
    discounted cash flow, 32–33
    net present value, 32–35
    project benefits, 30–31
    project success, 30–31
    return on assets, 32
    return on investment, 32
Financial strategies, 21f, 21t
First article inspection (FAI), 205
First in, first out (FIFO), 69, 205
First-pass yield (FPY), 205
Fishbone diagram, 109, 198f, 205
5S, 121, 203
Five forces analysis, 181–183, 183f
Five whys, 116–117, 203
Flowchart, 70
Flow/flow production, 70, 205

G

Galilee, Galileo, 72
Galvin, Bob, 1
Gantt, Henry, 60
Gantt chart, 60–62, 61f
Gap analysis, 115–116
Gemba/gemba walk, 205
Global Quality Management System
    (GQMS), 201
Goldratt, Eliot, 121
Green Belt, 165

H

Harry, Mikel, 1
Heijunka, 206
Histograms, 53
Hoshin Kanri, 206
House of quality, 40–51, 40f, 42f–49f
Hypothesis test definitions, 113–114
Hypothesis testing, 112–115

I

Improve phase
    5S, 121
    business process improvement, 119
    cause-and-effect matrix, 119–120, 120f

continuous flow manufacturing, 121
continuous improvement, 34, 125, 158
error proofing, 121
failure mode and effects analysis, 117, 119
Lean tools, 121, 122f
plan–do–check–act process, 119
prioritization process, 119–120, 120f
process ID, 119–120
pull, 121
setup reduction, 121
SIPOC, 119
standard work, 121
theory of constraints, 119, 121–123
total productive maintenance, 121
value stream mapping, 121
Improvements, sustenance of, 156–159
Inflection point, 10
Internal business strategies, 21f, 21t
Internal setup, 206
Ishikawa diagram, 206
ISO TS 17349, 74

J

Jidoka, 206
Just in time (JIT), 206

K

Kaikaku, 206
Kaizen approach, 28, 119, 201, 206
Kaizen event, 206
Kanban, 206
Key critical characteristic (KCC), 206
Key performance characteristic (KPC), 206
Key performance indicators (KPIs), 20–21,
    206

L

Lead time, 206
Lean Six Sigma (LSS); *see also* Six Sigma
    analyze phase, 196–199, 198f, 199f
    case study, 187–201
    control phase, 200
    define phase, 192–195, 195f
    improve phase, 199–201, 200t, 201t
    improve process, 121, 122f
    integration and, 201

measure phase, 196, 197t
    methodologies for, 187–201
    objectives of, 191–192
    standardization and, 201
    strategies for, 187–191, 188f–190f,
        188t, 191t
    tools for, 121, 122f, 191–192
Learning and growth strategies, 21f, 21t
Levitt, Theodore C., 182
Line balancing, 206
Lower control limit (LCL) values, 83–84, 85f
Lower specification limits (LSL), 101–102

M

Malcolm Baldrige National Quality
    Award (MBNQA), 207
Management system, 16
Manufacturing Requirement Planning
    (MRP), 206
Manufacturing Resource Planning
    (MRP II), 206
Marketing, 181–182
Mean, 8–12, 12f
Mean time between failures (MTBF),
    172–173
Mean time to failure (MTTF), 172–173
Measure, analyze, improve, and control
    (MAIC) approach, 1; *see also*
    DMAIC process
Measure phase
    definitions, 77
    measurement equipment, 76–77
    measurement system analysis, 74–98
    measurement system process, 80–98
    process capability measurement, 98–105
    process characteristics, 67–74
Measurement scales, 73–74, 74t
Measurement system analysis (MSA)
    accuracy, 78–79, 78f
    appraiser variation, 79–80
    bias, 77–78, 77f
    characteristics of, 78–79
    components of, 78–80
    definitions, 77
    degree of accuracy, 78–79, 78f
    equipment variation, 79–80
    errors, 79–80, 80f
    measurement equipment, 76–77

measurement system process, 80–81
precision, 79–80
sampling strategies, 75–76, 76t
system errors, 79–80, 80f
terms, 77
uses of, 74–76
Measurement system process
accuracy determination, 88–89
analysis of, 92–93, 93t
confidence levels, 92–95, 93t
control chart constants, 82–83, 84t, 95
data sheet, 94, 95f
linearity data, 89–90, 90t, 91f
linearity evaluation, 89–90
linearity graph, 91, 91f
lower control limit values, 83–84, 85f
mean charts, 83, 86f
measurement order, 91–92, 92t
out-of-control charts, 84–86, 86f
preparing for study, 81–82
range charts, 82–83, 84t, 86f
repeatability, 91–94, 95f, 96f, 97f
reproducibility, 91–94, 95f, 96f, 97f
resolution evaluation, 87–88
stability evaluation, 82–83, 83f
upper control limit values, 83–84, 85f
Median, 8, 10, 11f, 12, 12f
Metric, 12–16
Milk run, 207
Mistake proofing, 121, 166, 168
Mode, 8, 10, 11f, 12, 12f
Motorola, 1–2
Muda, 207
Mura, 207
Muri, 207

N

Net present value (NPV), 32–35
Net profit (NP), 123
Normal distribution, 9–11, 9f, 11f, 12f, 13–15, 14f
Null hypothesis, 113; *see also* Hypothesis testing

O

One-piece flow, 207
Operational excellence (Opex), 207

Organizational goals, 123
Organizational performance measures
balanced scorecard, 20, 21f
cost–benefit analysis, 30–32
critical-to-quality measures, 21
effectiveness of, 20f
financial measures, 30–35
key performance indicators, 20–21
net present value, 32–35
quality performance measures, 22–30
return on assets, 32
return on investment, 32
stakeholders, 19–20
Organizational processes
balanced scorecard, 20, 21f, 22f
customer strategies, 21f, 21t
effectiveness of, 20f
feedback loop, 20, 20f
financial strategies, 21f, 21t
internal business strategies, 21f, 21t
learning and growth strategies, 21f, 21t
stakeholders, 19–20
strategies, 20–21, 21f, 21t
team management, 35
Out-of-control charts, 84–86, 86f
Overall equipment effectiveness (OEE), 207

P

Pareto chart, 105, 107, 199, 199f
Part dimensional specifications, 177–178, 178f
Parts per million (ppm), 15, 16f, 24
Percentage rates, 33–34
Performance measures, 29–35; *see also* Business performance measures
PERT chart, 63–66, 64f
Plan–do–check–act (PDCA) process, 28–30, 119, 201, 207
Point-of-use storage (POUS), 207
Poisson model, 152
Poka Yoke, 166, 207
Population parameters, 8–11, 9f
Porter, Michael, 182
Pp/Ppk, 100–103
Precision components, 79–80, 79f
Procedures, 70
Process capability measurement

assessments, 100–103, 101f, 103t
diagrams, 98, 99t, 100f
example of, 98, 99t
height diagrams, 98, 99t
histograms, 98, 100f, 101f
long-term Sigma, 102–103, 103t
process capability studies, 104–106, 159
short-term Sigma, 102–103, 103t
Process capability studies, 104–110, 159
Process characteristics
cross-functional areas, 72
data collection, 72–73
input/output variables, 67–70
measurement scales, 73–74, 74t
process handoff diagram, 70–72, 71f
process model, 68f
quality system documents, 68f
Process control plan (PCP), 207
Process definition, 1–2, 207–208
Process failure mode and effects analysis (PFMEA), 207
Process for performance analysis and improvement (PAIP), 207
Process handoff diagram, 70–72, 71f
Process ID, 119–120
Process map, 70
Process metrics, 157, 159, 201
Process spread, 3–4
Process stability, 154, 155
Process variability, 3–4, 5f, 77, 85, 121, 136, 157
Product life cycle cost, 163, 163f
Product realization, 208
Product spread, 3–4
Productivity, 17, 123, 208
Program evaluation and review technique (PERT), 37, 63–66, 64f
Project charter, 53–60, 58f
Project goal, 54
Project performance, 56, 57f, 58f, 59f
Project scope, 54
Project statement, 53–54, 56t
Project status report, 59f
Project tracking, 60–63
Psychological inertia, 184, 184f
Pugh matrix, 119
Pull, 121, 208
Push, 208

Q

Qualitative data, 73
Quality, house of, 40–51, 40f, 42f–49f
Quality function deployment (QFD)
absolute importance, 49f
competitor products analysis, 46f
competitor rating by customers, 43f
correlation matrix, 41, 44, 48f
customer requirements, 43f, 50t
expanded house of quality, 41–51, 42f–49f
explanation of, 4
house of quality, 40–51, 40f, 42f–49f
meeting needs of customers, 44f, 45f, 50t
relative importance, 49f
target values, 47f
technical specifications, 47f
Quality operating system (QOS), 208
Quality performance measures
cost of poor quality, 24, 27f
defects per million opportunities, 23–24
defects per unit, 22–30
parts per million, 15, 16f, 24
rolled throughput yield, 23–24
throughput yield, 23
Quantitative data, 73

R

Range charts, 82–83, 84t, 86f
Regression analysis, 90, 108–109
Reliability
accelerated failures, 170, 174
bathtub curve, 170–171, 170f, 174f
burn-in and, 170–171, 170f
customer concerns and, 168–171
Design for Six Sigma and, 168–177
failure rates and, 170–177, 174f
mean time between failures and, 172–173
mean time to failure and, 172–173
operating conditions and, 170
probability and, 169, 172, 174–177
stable operating life, 171–177
successful performance and, 169
time and, 170
wear-out failures, 170, 174–177, 174f

Repeatability, 78–80, 79f, 91–94, 95f, 96f, 97f
Reproducibility, 78–80, 79f, 91–94, 95f, 96f, 97f
Return on assets (ROA), 32
Return on investment (ROI), 32, 123, 162–163
Risk priority number (RPN), 208
Rolled throughput yield (RTY), 23–24, 208
Root cause analysis, 116–117

S

Scatter diagrams, 53, 108–109, 109f
Setup reduction, 121
7 waste, 203
Shadow board, 208
Shewhart, Walter, 125, 127
Significant characteristic description, 154
Significant characteristic number, 154
Single-minute exchange of dies (SMED), 208
Single-piece flow, 207
SIPOC (suppliers, inputs, process, outputs, and customers), 67–69, 119, 208
Six Sigma; *see also* Lean Six Sigma
    basic statistics, 7–8
    case study, 187–201
    central limit theorem and, 9–10, 9f
    central tendency and, 11–12
    defect assessment levels, 15–17, 15f, 16f, 16t
    define phase, 37–66
    definition of, 208
    design for, 161–185, 162f
    diagram of, 162f
    evolution of, 1–2
    explanation of, 7–8, 12–13
    for improvement process, 121, 122f
    interpretation of, 13
    introduction to, 1–2
    Lean tools, 121, 122f
    as management system, 16
    measure of dispersion and, 12–13, 13f
    as methodology, 17, 19
    as metric, 12–16

normal distribution and, 9–11, 9f, 11f, 12f, 13–15, 14f
    origin of, 1–2
    population parameters, 9–10, 9f
    process of, 4, 5f
    sigma levels, 15–17, 15f, 16f, 16t
    statistical interpretation of, 13
    tools for, 162
    understanding, 7–8
Slack time, 65
SMART goals, 56
Smith, Bill, 1
Spaghetti diagram, 69, 70, 70f, 208
Stability evaluation, 82–83, 83f, 159
Stakeholders, 19–20
Standard normal distribution, 9–11, 9f, 11f, 12f, 13–15, 14f
Standard operating procedures (SOPs), 157–158
Standard work, 121, 157, 208
Standard work instructions, 70, 157–158
Statistical process control (SPC)
    background on, 125
    benefits of, 126
    capability and, 126
    common causes and, 127, 127f
    concepts of, 126–128
    definition of, 208
    explanation of, 4, 125
    objectives of, 126
    overadjustments and, 126
    special causes and, 127–128, 128f
    stability and, 126
    uses of, 126
    variation and, 126–128, 127f, 128f
Statistics
    analytical statistics, 7–8
    basic statistics, 7–8
    central limit theorem for, 9–10, 9f
    definition of, 8–9
    descriptive statistics, 7
    enumerative statistics, 7
    explanation of, 8–9
    inferential statistics, 7
    measurement of, 7–8
    normal distribution and, 9–11, 9f
    population parameters and, 8–9
    types of, 7–8

Stratification technique, 53
Strength, weakness, opportunity, and threat (SWOT), 209
Supermarket, 209
Sustainability, 209
Symmetrical normal distribution, 10–11, 11f
System definition, 2, 209

T

Take rate, 159
Takt time, 69, 209
Target (T) value, 101–104
Team management, 35
Theory of constraints (TOC), 119, 121–123
Throughput, 70
Throughput yield (TY), 23
Time metrics, 59
Tolerance
    Design for Six Sigma and, 177–181, 178f
    intervals, 178, 179t
    specifications for, 177–178, 178f
    stack tolerance, 178–179, 180f
    statistical tolerance, 177–178
    statistical tolerancing, 180–181
Total productive maintenance (TPM), 121, 209
Total quality management (TQM), 209
Touch time, 67–69
Training plan, 156–157
TRIZ approach, 183–185, 185t
Turnover, 123

U

Upper control limit (UCL) values, 83–84, 85f
Upper specification limits (USL), 101–102

V

Value definition, 37, 209
Value stream, 69–70, 209
Value stream map (VSM), 70, 121
Value-added activity (VAA), 209
Variable data charts, 130–140; *see also* Control charts
Variance, 8, 9
Variation
    definition of, 1
    measuring, 1–3, 3f
    role of, 1
    spread of, 3–4
    types of, 1–4, 2f, 3f
Visual factory, 158, 210
Voice of customer (VOC)
    customer satisfaction, 38–39
    customer–supplier chain, 38, 39f
    external customers, 38
    internal customers, 38
    satisfaction survey for, 39

W

Waste analysis, 107
Work in progress (WIP), 67–69, 159, 210
Work in queue (WIQ), 67–69

About the Author

 **Suresh Patel** is a former Technical Director and Operations Excellence Executive. He holds a BE degree in electrical engineering from M.S. University of Baroda, India, a Master's degree in production technology from South Bank University, London, and an MBA degree from the University of Texas at Brownsville. He is qualified as a Certified Reliability Engineer, Certified Quality Engineer, and Certified Management Systems Auditor by the American Society for Quality.

In his long career spanning more than four decades, he has developed a wide range of products/processes and has helped in establishing six manufacturing plants in India and five U.S. plants in Mexico. Starting with India, his career path has enabled him to work in industries in the UK, Denmark, Belgium, Canada, U.S., China, Mexico, and Chile. His career has been enriched through holding key positions with such multinational corporations as Gestetner, Motorola, United Lighting Technologies, Eaton Corporation, and Fiat Global.

Patel's practical expertise and interests include establishing Business Excellence Strategies starting from product quality strategies, quality improvement tools deployment, and execution, leading to improvements in product/process delivery performance and reduction in product escapes and product/process variation through Lean Six Sigma and overall business excellence employing Leadership and Results "Triades" as defined in MBNQA USA. Patel's other interests include supply chain management, manufacturing management, and building technological capabilities in manufacturing firms.